Differentiated Instruction for Mathematics

Instructions and activities for the diverse classroom

Hope Martin

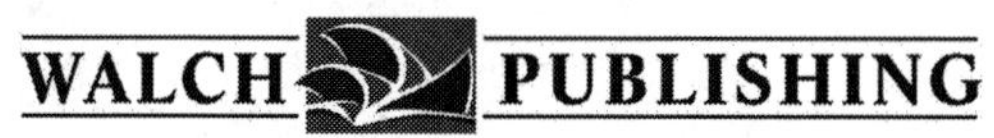

1 2 3 4 5 6 7 8 9 10
ISBN 0-8251-5895-8

J. Weston Walch, Publisher
P.O. Box 658 • Portland, Maine 04104-0658
Printed in the United States of America

Table of Contents

Introduction

Mathematics is the key to opportunity. . . . For students, it opens doors to careers. For citizens, it enables informed decisions. For nations, it provides knowledge to compete in a technological economy.

—NATIONAL RESEARCH COUNCIL (1989)

To meet the needs of all students and design programs that are responsive to the intellectual strengths and personal interests of students, we must explore alternatives to traditional mathematics instruction. We need to examine not only what is taught but how it is taught and how students learn.

Carol Ann Tomlinson in *The Differentiated Classroom: Responding to the Needs of All Learners* encourages educators to look at teaching and learning in a new way. Using the phrase "One size doesn't fit all," she presents, not a recipe for teaching, but a philosophy of educational beliefs:

- Students must be seen as individuals. While students are assigned grade levels by age, they differ in their readiness to learn, their interests, and their style of learning.
- These differences are significant enough to require teachers to make accommodations and differentiate by content, process, and student products. Curriculum tells us what to teach; differentiation gives us strategies to make teaching more successful.
- Students learn best when connections are made between the curriculum, student interests, and students' previous learning experiences.
- Students should be given the opportunity to work in flexible groups. Different lessons point toward grouping students in different ways: individually, heterogeneously, homogeneously, in a whole group, by student interests, and so forth.
- There should be on-going assessment—assessment can be used to help plan effective instruction.

To address the diverse ways that students learn and their learning styles, we can look to Howard Gardner's eight intelligences to provide a framework. Gardner's theory of multiple intelligences encourages us to scrutinize our attitudes toward mathematical learning so that each student can learn in a more relaxed environment.

Let's explore what multiple intelligences look like in the mathematics classroom.

Visual/Spatial

Perceives the visual world with accuracy; can transform and visualize three dimensions in a two-dimensional space. Encourage this intelligence by using graphs and making sketches, exploring spatial visualization problems, relating patterns in math to visual and color patterns, using mapping activities, and using manipulatives to connect concrete with abstract.

Verbal/Linguistic

Appreciates and understands the structure, meaning, and function of language. These students can communicate effectively in both written and verbal form. Encourage this intelligence by using class to discuss mathematical ideas, using journals to explore mathematical ideas using words, making written and oral presentations, and doing research projects.

Logical/Mathematical

Ability to recognize logical or numerical patterns and observe patterns in symbolic form. Enjoys problems requiring the use of deductive or inductive reasoning and is able to follow a chain of reasoning. Encourage this intelligence by organizing and analyzing data, designing and working with spreadsheets, working on critical-thinking and estimation problems, and helping students make predictions based upon the analysis of numerical data.

Musical/Rhythmic

The ability to produce and/or appreciate rhythm and music. Students may enjoy listening to music, playing an instrument, writing music or lyrics, or moving to the rhythms associated with music. Activities related to this intelligence include using songs to illustrate math skills and/or concepts and connecting rational numbers to musical symbols, frequencies, and other real-world applications.

Bodily/Kinesthetic

The ability to handle one's body with skill and control, such as dancers, sports stars, and craftspeople. Students who excel in this intelligence are often hands-on learners. Activities related to this intelligence include the use of manipulatives, involvement with hands-on activities (weighing, measuring, building), and permitting students to participate in activities that require movement or relate physical movements to mathematical concepts.

Interpersonal

The ability to pick up on the feelings of others. Students who excel in this intelligence like to communicate, empathize, and socialize. Activities related to this intelligence include using cooperative-learning groups, brainstorming ideas, employing a creative use of grouping (including heterogeneous, homogeneous, self-directed, and so forth), and using long-range group projects.

Intrapersonal

Understanding and being in touch with one's feelings is at the center of this intelligence. Activities related to this intelligence include encouraging students to be self-reflective and explain their reasoning, using journal questions to support metacognition, and giving students quiet time to work independently.

Naturalist

Naturalist intelligence deals with sensing patterns in and making connections to elements in nature. These students often like to collect, classify, or read about things from nature—rocks, fossils, butterflies, feathers, shells, and the like. Activities related to this intelligence include classifying objects based upon their commonalities, searching for patterns, and using Venn diagrams to help organize data.

The Format of the Book

The National Council of Teachers of Mathematics (NCTM) in *Principles and Standards for School Mathematics* (2000) refined the 1989 standards by delineating content and process goals essential for all students, grades K–12. The chapters of this book have been organized around the content and process standards defined by the NCTM—numbers and operations, algebra, geometry and measurement, and data analysis and probability. The activities and projects in each chapter reflect the philosophy of differentiation, provide a math curriculum that is standards-based, and involve students in hands-on, motivating real-world problems. Each chapter ends with a "Brush Up Those Skills" game that supports flexible grouping and employs the skills and concepts introduced in the chapter's activities. The Appendix contains copies of a blank Teacher's Page, Task/Audience/Product (TAP) Activities page, and "Brush Up Those Skills" pages to help the teacher design and organize his or her own differentiated mathematics lessons. One of the lessons, "How Long Is Your First Name?" requires students to use one-inch squares, and a "Brush Up Those Skills" activity requires a copy of Pascal's triangle—so these sheets have been provided, as well. The Teacher's Pages require additional discussion.

Using the Teacher's Pages

Each mathematical experience is preceded by a Teacher's Page that includes valuable information for managing the lesson. These pages have been designed to merge the NCTM's mathematics standards and the philosophy of differentiation.

- **Math Topics:** As in most hands-on activities, these mathematical experiences address more than one math skill or topic. In the real world, mathematics is an integrated experience, and skills and concepts interrelate and blend. When using the activities, teachers can use this section to connect the lesson to skills and concepts that are part of their mathematics curriculum.

- **Prior Knowledge Needed:** Differentiating necessitates that the teacher know their students' prior knowledge. This information can be gained in a variety of ways—through pretesting, observation and questioning, or available data from other sources. To be assured that each student's experience is meaningful and enriching, it is important to know where to start.

- **Differentiation Strategies**

  Principles: The three principles of differentiation are respectful tasks, flexible grouping, and ongoing assessment. Applicable principles are discussed in this section.

  Teacher's Strategies: Teachers can differentiate content, process, and/or product. Every student should be exposed to a mathematics curriculum that is equitable, essential, and requires higher-level thinking skills. Those differentiation strategies that can be used to accommodate diverse learners can be found in this section.

  According to Students: Students' readiness, interests, and learning styles determine the accommodations made in a differentiated classroom. Pretest and interest surveys can be taken at any time to meet student needs. This section suggests the multiple intelligences and learning styles that are highlighted in the activity.

- **Materials Needed**

  A comprehensive list of materials and supplies needed for each activity is provided. To help the activity run more smoothly, these should be gathered and made ready prior to the lesson.

- **Teaching Suggestions**

  Engaging the Students: Suggestions are made to begin the lesson and perk student interest. There may be a song or a poem that will draw student attention. This engaging activity varies with the lesson. Productive questioning is sometimes suggested to focus students on the lesson. Productive questioning include these questions:

  - Focus attention (What have you noticed about . . . ? What do you see when you . . . ?)
  - Help students count or measure (How many . . . ? How long . . . ? How much . . . ?)
  - Comparison questions (What do these have in common with . . . ? How are they different?)

- Problem-posing questions (What problems did you face when . . . ? How did you solve this problem?)
- Reasoning questions (Why do you think . . . ? What is your reason for . . . ? Can you come up with a rule for . . . ?)

These types of questions can also be used during debriefing or as ongoing assessment during observation of students.

The Exploration: "Engaging the Students" usually has students discussing and working as a whole group. If students are going to work individually, with a partner or in a group of four, the regrouping will be done at this time. While this section is not a scripted, step-by-step plan, it does give suggestions to help teachers encourage students and make the experience more meaningful. Some lessons do suggest specific questions that will help students focus or develop understanding. But these are merely suggestions and should be used only if appropriate to the needs of the class. Available answers may also appear in this section. Other selected answers are found at the back of the book.

Debriefing: During this time groups, will come back together to discuss their findings and share their results. Many of the activities ask students to share their data on a group data table for further analysis. Productive questions can be used at this time to help focus students' attention on the important concepts and skills presented in the activity.

- **Assessment**

Multiple suggestions are made here. They may include the following: (1) completed student products, (2) observation and questioning, (3) tiered or non-tiered journal question(s), and (4) TAP activities.

Many types of activities can be tiered, such as assignments, journal questions, warm-ups, and activities. A few suggestions to help plan tiered activities are as follows:

1. All levels should apply the same skill or concept.
2. All levels should meet student readiness.
3. All levels should challenge students and provide new learning experiences.
4. Higher difficulty levels should be a more faceted problem, have less structure, and require more independence of the students.
5. All levels should build on student knowledge and require higher level thinking skills.

Task/Audience/Product (TAP) activities are adapted from Tomlinson's RAFT activities. They are used sparingly in assessment suggestions, but a blank form is located in the Appendix section of this book. This form can be used to develop your own assessment activities.

- **Variations for Differentiation (Tiered Activities):** Some of the activities can be tiered or elaborated upon, and suggestions have been made in this section. If the teacher has any activities that can be added to enhance the lesson, this would be a fine place in which to list those for future use.

Chart of Multiple Intelligences

Mathematics Strands	Activities	Logical/Mathematical	Verbal/Linguistic	Bodily/Kinesthetic	Musical/Rhythmic	Visual/Spatial	Interpersonal	Intrapersonal	Naturalist
Number Theory, Numeration, and Computation	The Bowling Game	●	●		●		●		
	All About the Moon	●	●				●		
	The Pattern Tells It All	●	●		●	●		●	●
	Baking Blueberry Muffins	●	●	●	●		●		
	Eratosthenes and the 500 Chart	●	●		●	●		●	●
	Diagramming Divisibility	●					●	●	●
Patterns, Functions, and Algebra	The Parachute Jump	●	●	●	●	●	●		
	The Irrational Spiral	●		●	●	●	●		
	Seeing to the Horizon	●	●				●		
	Falling Objects: Formulas in the Real World	●	●	●			●		
	Pattern Block Patterns	●	●	●		●	●		●
Measurement and Geometry	Pennies and the Sears Tower	●	●	●		●	●		
	The Bouncing Ball	●	●	●			●	●	
	Soda Pop Math	●	●	●		●	●		
	Paper-Folding Polygons	●	●	●		●		●	●
	Graphing the Area of a Rectangle	●	●			●		●	●
	The Valley on Mars	●	●	●		●	●		
Data Analysis, Statistics, and Probability	Predicting Colors in a Bag of M&M's	●	●	●			●		
	Vowels, Vowels, Everywhere	●	●		●	●	●		
	Phone Home	●	●	●			●	●	
	How Long Is Your First Name?	●	●	●			●	●	
	Geometric Probability: Dartboards and Spinners	●	●	●		●	●	●	
	Flipping Three Coins: Heads or Tails?	●	●	●		●	●		

Number Theory, Numeration, and Computation

There is nothing so troublesome to mathematical practice . . . than multiplications, divisions, square and cubical extractions of great numbers . . . I began therefore to consider . . . how I might remove those hindrances.

—JOHN NAPIER

Number theory, numeration, and computation remain important components of the current school mathematics curriculum. While it is true that computation and basic number facts have been emphasized to the detriment of other strands of mathematics, we all understand that student proficiency in these areas is essential for students to be successful in understanding math concepts. With guidance and meaningful experiences, students will gain a sense of number, improve their ability to solve problems, and develop useful strategies to estimate reasonableness of answers. The activities and projects in this chapter will encourage the development of number and operation sense in students.

"The Bowling Game" (page 3) gives students the opportunity to practice order of operations and computational skills in an interesting way. The numbers used in the game are determined by chance according to the roll of one die. Individual creativity and imagination are rewarded in points gained. The lesson can be augmented by using a little ditty, "O^3—Order of Operations Song" (page 6). This makes learning the rules even more fun and draws attention to the musical/rhythmic intelligence!

"All About the Moon" (page 8) demonstrates the connections between math and science while presenting computation, estimation, conversions, and open-ended problem solving as captivating activities. Interesting moon facts are explained in a way that makes sense to students, and very large numbers become more understandable through hands-on activities. By encouraging students to work collaboratively, a variety of differentiation strategies can be utilized.

When patterns found in numbers are translated into fascinating visual patterns, students can investigate the relationship between art and mathematics; skills and concepts are learned using a variety of learning styles and intelligences. In "The Pattern Tells It All" (page 12), students learn how prime and composite numbers form patterns in number grids that are visually different. By writing about their patterns, students verbalize the depth of their understanding.

When students can see connections between school math and real math, they are motivated to learn important concepts and skills. "Baking Blueberry Muffins" (page 18) shows students mathematics applications that utilize ratio and proportion, percent of profit, and everyday conversions. While students work together on this activity, they make use of many of the strategies of differentiated instruction.

By using the rules of divisibility and some simple computation, students can discover all of the prime numbers less than 500 using "Eratosthenes and the 500 Chart" (page 21). "The Divisibility Ditty" (page 115) reminds students of the divisibility rules while encouraging students to use a variety of intelligences to learn important mathematics concepts.

Finally, "Diagramming Divisibility" introduces Venn diagrams—using logic to reinforce divisibility rules in yet another way. "The Divisibility Ditty" can be used again to review these important rules.

The Bowling Game

MATH TOPICS

whole number computation, order of operations, problem solving

PRIOR KNOWLEDGE NEEDED

1. whole number facts
2. order of operations—Parentheses, Exponents, Division, Multiplication, Add, Subtract (PEDMAS)

DIFFERENTIATION STRATEGIES

Principles

Flexible grouping: Students work in pairs or in groups of four to play the Bowling Game. Students can be paired either homogeneously or heterogeneously (based upon using the tiered activity discussed below).

Ongoing assessment: Use student game sheets to assess level of achievement.

Teacher's Strategies

Product: Tiered Journal Questions: Based upon student readiness, journal question Level 1 or journal question Level 2 can be assigned. Level 1 is easier because it contains only three numbers, not four, but each problem requires students to use order of operations. Be sure to have students share their solutions with the class as these problems are open-ended—they have one correct answer but there is more than one way to get to that answer.

A Tiered Activity: All students can participate in this activity by using their four numbers to find the numbers 1–10. Students with more advanced skills can be encouraged to use the same numbers and find the numbers 1–20.

According to Students

Learning Styles/Multiple Intelligences: musical, logical/mathematical, interpersonal, and verbal/linguistic

MATERIALS NEEDED

1. one game sheet for each player for each game
2. one die for each player
3. a timer or watch with second hand

4. copies of "O^3—Order of Operations Song"

TEACHING SUGGESTIONS

Engaging the Students

Begin the lesson by having students sing the song on the order of operations. Ask students, "Based upon this song, why do we say the order of operations can be defined by the letters in the mnemonic device PEDMAS? Review of order of operations can be accomplished using the song and this mnemonic device. These strategies help meet the learning needs of students who excel in musical intelligence or are verbal/linguistic intelligences.

The Exploration

Students can play in pairs or in groups of four. However, each player will need his or her own game board. Students will need one board for each game they play. A maximum of five minutes has been allotted for each game. However, the teacher may change the time to suit the needs and skill level of the students. Discuss these rules:

1. Each player rolls one die four times and enters the numbers in the space provided. Each student plays with his or her own set of four numbers.
2. After each player has acquired numbers, the timing begins. Each player has five minutes to find as many of the numbers (1–10) as possible with the four numbers. Students can add, subtract, multiply, divide, use exponents, parentheses, square roots, or any combination of the above—but they must use the order of operations.
3. A player receives one point for each number found, but if all ten numbers are found, the player receives 20 points (this is a strike).
4. When time is up, the teams can play another game. Each player gets a new game board and rolls the die again. The player will have a new set of four numbers to work with.
5. Play continues until one of the players has 30 points or until time has expired for the round of play.

Students are given space on the game sheet to record their solutions. These sheets should be collected and can be used as part of the assessment. It is important that students record their solutions so that their understanding can be evaluated.

Debriefing

At this time, students will be asked to share their strategies and methods of solutions. Connections can be made between student strategies and order of operation concepts.

ASSESSMENT

1. Student products: Game sheets can be collected and used to determine level of learning for each student.
2. Journal questions:
   a. Level 1 question: This problem can be solved in many ways. Use order of operations to help you find the answer (12). You can add, subtract, multiply, divide, and use parentheses, exponents, or square roots. The numbers can be used in any order.

   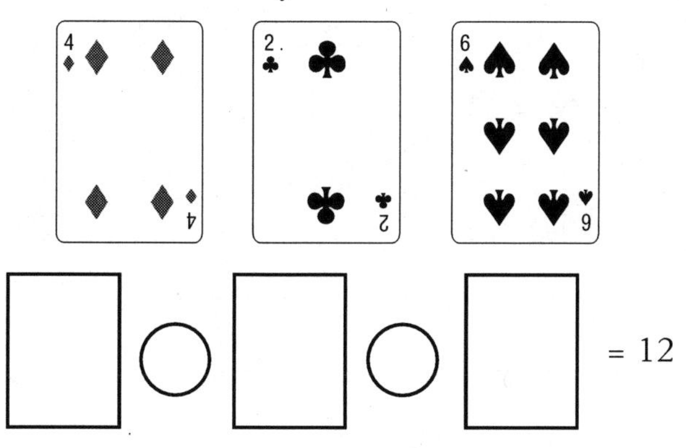

   Be sure to explain how you solved this problem.

   b. Level 2 question: This problem can be solved in many ways. Use order of operations to help you find the answer (0). You can add, subtract, multiply, divide, and use parentheses, exponents, or square roots. The numbers can be used in any order.

   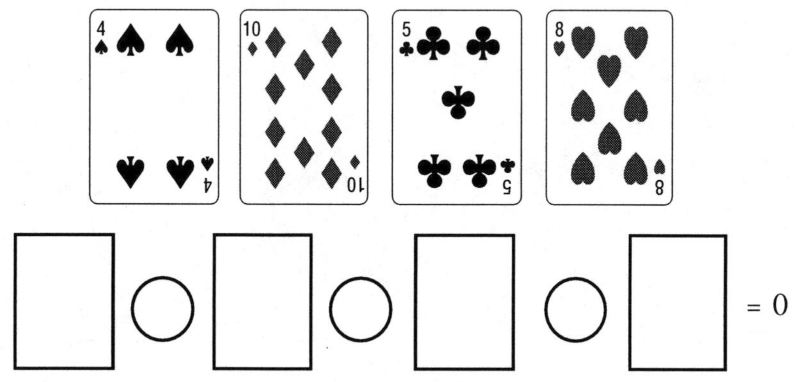

   Be sure to explain how you solved this problem.

O^3—Order of Operations Song

SUNG TO "FRÈRE JACQUES"

Do a problem, in what order?

What to do? There are rules.

First we do parentheses, they're the first to look at,

We're no fools, use the rules.

Next exponents, roots, and powers,

Once that's done, what comes next?

One step for each lesson, now we're on a mission,

We're no fools, use the rules.

Mul-ti-pli-ca-tion or di-vi-sion,

They come next, left to right,

Adding and subtracting, ends our little problem.

"O" to the third, we've been heard!

The Bowling Game

Directions: Roll one die four times. Write the numbers in the spaces below. Use the order of operations to find each of the numbers (1–10) to "knock down" each bowling pin. If you can find all ten numbers, you have a strike and get 20 points. Otherwise you get 1 point for each pin you "knock down." The game continues for five minutes or until neither player can find any more numbers (whichever comes first). Then each player rolls the die again and begins a new game. The first player with 30 points wins. Be sure to show your work for each game in the space provided.

7 8 9 10
4 5 6
2 3
1

My Numbers

My Solutions

All About the Moon

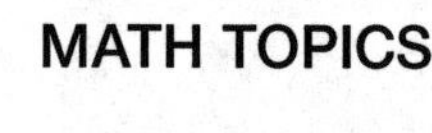

MATH TOPICS

computation, rounding, open-ended problem solving, and conversions

PRIOR KNOWLEDGE NEEDED

1. converting hours to years
2. converting pounds to ounces
3. rounding

DIFFERENTIATION STRATEGIES

Principles

Flexible grouping: Students work in pairs to solve these problems. Grouping can be homogeneous or heterogeneous.

Ongoing assessment: Students are asked to show their work and explain how they solved the problems. These can be used for assessment.

Teacher's Strategies

Product: Tiered Journal Questions: Based on student readiness, journal question Level 1 or journal question Level 2 can be assigned. Level 1 requires conversions but since the numbers are smaller, it is an easier problem to solve than Level 2.

According to Students

Learning Styles/Multiple Intelligences: logical/mathematical, verbal/linguistic, interpersonal

MATERIALS NEEDED

1. an activity sheet for each pair of students
2. calculators
3. a world atlas or access to Internet geography sites

TEACHING SUGGESTIONS

Engaging the Students

Ask students, "What were you doing 239,000 seconds ago?" When they take out calculators or a pencil and paper ask, "Why don't you know right away what you were doing? Do you know what you were doing one hour ago? Why?" After some discussion, students will realize that because 239,000 is a large number and a second is so small, it is probably not the best unit to describe this length of time (239,000 seconds $\approx 2\frac{3}{4}$ days). Explain to students that this activity describes some very interesting facts about the moon but asks us to try to make sense of these enormous numbers by comparing them to things we do in our everyday lives. In order for these numbers to make sense, they need to be labeled in the most appropriate units—just like $2\frac{3}{4}$ days makes more sense to us than 239,000 seconds.

The Exploration

Working in pairs, students are asked to do the following:

1. Using reference materials, compare the size of a crater on the moon to countries that might fit into it. A possible answer might be Lithuania, which is 826 square miles smaller than the crater.
2. Calculate how long it would take a train traveling 75 mph to get to the moon (239,000 miles away). Using calculators, students find that it takes 3186.67 hours to get to the moon. But is hours an appropriate unit to use? How long a time is this? 3186.67 hours is approximately 133 days.
3. The Apollo Space Program has cost about $25.5 billion dollars! Do we really understand the significance of a number this large? Students calculate that each pound of moon rock costs $30 million or $1,875,000 per ounce. Supposedly, the most expensive perfume is Joy, and it sells for about $65 per ounce.

Debriefing

These problems are open-ended, because there is more than one way to find the answer and, in some instances, more than one correct answer. It is important to give students an opportunity to explain, not only what answers they got, but how they solved the problem. By doing this, you are saying the process (the thinking) is as valuable as the product (the answer).

ASSESSMENT

1. Student products: Student answers to problems posed on the activity sheet can be used for assessment.
2. Extension activities:
   a. Groups of students can research the density of populations and based on these ratios, calculate how many people might live in a crater the size of Bailly Crater.
   b. Students can research the cost per ounce of everyday objects and liquids such as a pair of sneakers, gasoline, their math book, and so forth.
3. Journal questions:
   a. Level 1 question: It is 1,248,000 inches from Phoenix, Arizona to the Grand Canyon. While this information is correct, what unit of measure would be more reasonable than inches? Express the distance from Phoenix to the Grand Canyon using this more reasonable unit of measure. Explain how you solved this problem.
   b. Level 2 question: You are told that someone traveled 31,680,000 inches from their home to New York. What unit of measure would be more reasonable than inches to describe this distance? Express the distance to New York using this more reasonable unit of measure. Explain how you solved this problem.

All About the Moon

1. Did you know?

The largest crater we can see on the moon is called Bailly and covers an area of about 26,000 square miles.

Problem

Use a world atlas to find what countries could fit into a crater of this size.

2. Did you know?

The average distance from Earth to the moon is 239,000 miles.

Problem

If you were traveling on a train going at a speed of 75 mph, how long would it take you to get to the moon? (Be sure to use appropriate units of measure.)

3. Did you know?

The Apollo Space Program has cost about $25,500,000,000. The Apollo astronauts who landed on the moon brought about 850 pounds of rocks back to Earth.

Problem

About how much did the United States spend for each pound of moon rock?

for each ounce of moon rock?

The Pattern Tells It All

MATH TOPICS

multiples, patterns, critical thinking, problem solving

PRIOR KNOWLEDGE NEEDED

1. multiplication and division facts
2. divisibility rules for 2, 3, 4, 5, 6, 8, and 9

DIFFERENTIATION STRATEGIES

Principles

Flexible grouping: For this activity, students will work alone to shade in the multiple of each of the numbers.

Ongoing assessment: As students complete their designs, the designs can be used to assess understanding.

Teacher's Strategies

Product: TAP Activity: Students can choose one of the TAP activities; their products can be used for assessment.

According to Students

Learning Styles/Multiple Intelligences: visual/spatial, logical/mathematical, intrapersonal, verbal/linguistic, musical/rhythmic, naturalistic

MATERIALS NEEDED

1. If students will be required to shade in all of the multiples listed above, they will each need eight grids.
2. colored pencils or markers
3. copies of the song "The Divisibility Ditty"

TEACHING SUGGESTIONS

Engaging the Students

Ask students, "Do you think there might be a relationship between a pattern that we see in numbers and a visual pattern? Have you ever seen a visual pattern that seems to have a basis in mathematics?" Give students a chance to discuss this question and give any examples of patterns such as these. Then proceed with the exploration.

The Exploration

Give each student a copy of the number pattern sheet "The Pattern Tells It All" and ask him or her how it differs from the usual 100's chart. Students should notice that the first row and column have the multiples of one, the second row and column have the multiples of two, and so forth so that numbers are repeated throughout the chart. On the 100's chart, shading in the even number columns can identify the multiples of 2. On this chart, even numbers can be found throughout, and students will need to search for them.

Debriefing

The grid below shows the multiples of 3.

1	2	3	4	5	6	7	8	9	10
2	4	6	8	10	12	14	16	18	20
3	6	9	12	15	18	21	24	27	30
4	8	12	16	20	24	28	32	36	40
5	10	15	20	25	30	35	40	45	50
6	12	18	24	30	36	42	48	54	60
7	14	21	28	35	42	49	56	63	70
8	16	24	32	40	48	56	64	72	80
9	18	27	36	45	54	63	72	81	90
10	20	30	40	50	60	70	80	90	100

The grid below shows the multiples of 4.

1	2	3	4	5	6	7	8	9	10
2	4	6	8	10	12	14	16	18	20
3	6	9	12	15	18	21	24	27	30
4	8	12	16	20	24	28	32	36	40
5	10	15	20	25	30	35	40	45	50
6	12	18	24	30	36	42	48	54	60
7	14	21	28	35	42	49	56	63	70
8	16	24	32	40	48	56	64	72	80
9	18	27	36	45	54	63	72	81	90
10	20	30	40	50	60	70	80	90	100

Note: Each of the designs is a type of checkerboard, but the multiples of 4 have additional squares in the center of each larger square included in the design. As students color in their designs, they should begin to see this pattern emerge—the prime numbers are exact checkerboards and the composite numbers have additional factors (or shaded squares).

ASSESSMENT

1. Student products: Grids of multiple patterns from students can be used for assessment.
2. Journal question: Describe what the pattern might look like for the multiples of each of these numbers: 20 and 29. How would the patterns be alike? How would they be different? Why?
3. TAP Activities: Students will choose from one of the projects listed below:

Task	Audience	Product
poet	math students in your grade	Write a poem that will help students learn the rules for divisibility.
poster designer	students in your math class	Use the number grid provided in this lesson and design a poster that can be displayed in your math classroom that shows the patterns that result using the multiples of 2, 3, 4, 5, and 9.
board game designer	students in your math class	Design a board game that students can play to help them review both order of operations and divisibility rules.
author	students in third or fourth grade	Write a story about the number 5 and the number 10.
an original idea presented by student to be approved by the teacher		

The Divisibility Ditty

SUNG TO "THE ITSY BITSY SPIDER"

When looking for the factors
Of any number please,
Check out the rules
Then knowledge comes with ease.
Seek out only evens
Odds don't rate a look.
When dividing by two, my friend,
We'll go right by the book.

To work with three is fun,
Its rule is not too bad.
Sum up the digits
And see what sum you had.
If three goes into it,
It surely does you say,
Then the number that you started with
By three it was O.K.

Now four has other rules to watch
Don't look at odds, no way.
Check out the evens
And one more thing we say.
The last two digits,
Divisible by four.
That's the rule complete and true
We cannot tell you more.

Now what to say about the fives?
A fine and noble friend.
Fingers on one hand
And nickels we do spend.
Look out for fives and zeros
These numbers it must end.
Divisibility by five
Is certain my good friend.

Oh, six is odd and even too,
It takes two rules to solve.
Rules for two and three
This problem will resolve.
First it must be even,
And then divide by three,
Divide by six is possible,
It's mathematically!

Divide by eight is just like four,
The rule is much the same.
Check just the evens,
Dividing is the game.
Check the last three digits
Divisible by eight.
Another rule is in the bag,
Can we keep all this straight?

To solve for nine just look at three,
Their sameness they can't hide.
Just find the sum
By nine they must divide.
There's no remainder
These numbers do abound,
Fun with tunes and meter,
The rule for nines we've found.

The last to find—the number ten,
Its rule is short and sweet.
Zeros alone do
End these numbers neat.
We've learned all the rules of
Di-visi-bil-ity.
From 2 to 4 to 6 to 10
Start over now, oh gee!

The Pattern Tells It All

Directions: To complete this project, you will need eight number grids. On the first, shade in all the multiples of 2, on the second, shade the multiples of 3, then shade the multiples of 4, 5, 6, 7, 8, and 9. When you complete all eight designs, try to find a relationship between the number and the pattern it forms.

1	2	3	4	5	6	7	8	9	10
2	4	6	8	10	12	14	16	18	20
3	6	9	12	15	18	21	24	27	30
4	8	12	16	20	24	28	32	36	40
5	10	15	20	25	30	35	40	45	50
6	12	18	24	30	36	42	48	54	60
7	14	21	28	35	42	49	56	63	70
8	16	24	32	40	48	56	64	72	80
9	18	27	36	45	54	63	72	81	90
10	20	30	40	50	60	70	80	90	100

The Pattern Tells It All

Directions: Describe any relationships you see between the numbers used and the patterns they formed. (*Hint*: Think about what you know about prime and composite numbers.) Make a prediction—what do you think the pattern for the number 17 might look like? What about the pattern for the number 24? Explain your reasoning.

Baking Blueberry Muffins

Math Topics

computation, weight conversions, ratio and proportion, reading and analyzing data from a table, and problem solving

PRIOR KNOWLEDGE NEEDED

1. dividing (finding unit cost)
2. converting from mixed numbers to improper fractions
3. converting from ounces to pounds
4. finding volume
5. calculating percentage of profit

DIFFERENTIATION STRATEGIES

Principles

Flexible grouping: Students work in pairs to solve these problems.

Ongoing assessment: The student activity sheet requires students to read information from a table and answer six questions. These sheets can be collected and used to assess student progress.

Teacher's Strategies

Product: Tiered Journal Questions: Based on student readiness, journal question Level 1 or journal question Level 2 can be assigned. The Level 1 question is easier because its volume is a whole number (3 ounces) instead of a mixed number. Students will not be required to divide by a fraction to solve this problem. The Level 2 question contains unnecessary information and requires students to divide by a mixed number.

According to Students

Learning Styles/Multiple Intelligences: logical/mathematical, verbal/linguistic, bodily/kinesthetic, musical/rhythmic, interpersonal

MATERIALS NEEDED

1. one copy of the activity sheet for each pair of students
2. calculators

TEACHING SUGGESTIONS

Engaging the Students

Ask students, "If you went into a bakery to buy muffins, what would you do if all of the muffins were different sizes? Would you want to pay the same amount for a big muffin as you would for a small muffin? How do you think the baker makes muffins all the same size?" Discuss the need to use standard measures so that when professional bakers make muffins, the muffins are all the same size and weight. Explain to students that for this activity they will be professional pastry chefs and will be using the mathematics of cooking! (If you have scoops of different sizes, you can use these to show students the different volumes.)

The Exploration

Before students begin, have them explain the information provided on the table and problem-solve methods to convert ounces to fractions of a pound or pounds to ounces. (Since the batter is sold in 18-pound pails, the problem will require that the same units be used.) Allow students to work in pairs to solve the problems. Discuss the solutions when groups have completed their work.

Debriefing

Discuss how students solved the problems. Be sure to ask, "Did anyone solve this problem a different way?" It is important that students understand that process is as important as product.

ASSESSMENT

1. Student products: Solutions shown on activity sheet can be used to assess student progress.
2. Journal questions:
   a. Level 1 question: Cranberry muffin mix is sold in a 12-pound can. You will be using a #12 scoop. This size scoop is $\frac{1}{3}$ cup and holds 3 ounces of mix. How many muffins will you be able to make using a scoop this size?
   b. Level 2 question: A #30 scoop makes miniature muffins that have a volume of $2\frac{1}{5}$ tablespoons and weigh $1\frac{1}{4}$ ounces. How many mini-muffins can be made from a 12-pound can of mix?

Baking Blueberry Muffins

When professional bakers make muffins, scoops or dippers are used to control the portion size and to make sure that the muffins look alike and are the same size. Scoops are numbered, and the various sizes that appear on the lever indicate the number of level scoops it will take to fill a quart.

Number	Volume	Weight (oz)
6	$\frac{2}{3}$ cup	5
8	$\frac{1}{2}$ cup	4
10	$\frac{2}{5}$ cup	$3\frac{1}{2}$
12	$\frac{1}{3}$ cup	3
16	$\frac{1}{4}$ cup	$2\frac{1}{2}$
20	$3\frac{1}{5}$ tablespoon	$1\frac{2}{3}$
24	$2\frac{2}{3}$ tablespoon	$1\frac{1}{2}$
30	$2\frac{1}{5}$ tablespoon	$1\frac{1}{4}$
40	$1\frac{3}{4}$ tablespoon	1

If you were a food service professional, you would need to calculate how many portions you can get and how much you will charge for the muffins. Blueberry muffin batter is sold in 18-pound pails for $36.00.

Directions: Answer the following.

A. If you use a #12 scoop:
   1. How many muffins can you make?
   2. How much does each muffin cost?
   3. If you sell them for $18.00/dozen, what percent of profit are you making?

B. If you use a #6 scoop and make jumbo muffins:
   1. How many muffins can you make?
   2. How much does each muffin cost?
   3. If you want to make the same percent of profit, how much will you charge for each dozen?

Eratosthenes and the 500 Chart

Math Topics

prime and composite numbers, divisibility rules, computation, problem solving

PRIOR KNOWLEDGE NEEDED

1. definition of prime and composite numbers
2. divisibility rules
3. multiples

DIFFERENTIATION STRATEGIES

Principles

Flexible grouping: Each student will be designing his or her own chart of prime numbers.

Ongoing assessment: Assess successful completion of the 500 Chart.

Teacher's Strategies

Product: Tiered Journal Questions: Based upon student readiness, journal question Level 1 or journal question Level 2 can be assigned. Level 1 is easier because it asks students to discuss, in their own words, the procedures that were followed to produce the chart of prime numbers. Level 2 is more difficult because it asks students to attempt to discover a pattern. Since no pattern exists, the answer to this requires students to critically assess this lack of pattern to prime numbers.

According to Students

Learning Styles/Multiple Intelligences: logical/mathematical, visual/spatial, verbal/linguistic, musical/rhythmic, intrapersonal, naturalist

MATERIALS NEEDED

1. a copy of the 500 Chart for each student
2. highlighter for each student
3. calculators (using the song is preferable)
4. copies of the song "The Divisibility Ditty" on page 15

TEACHING SUGGESTIONS

Engaging the Students

A Greek mathematician, Eratosthenes, devised a procedure for finding prime numbers called the sieve of Eratosthenes. Ask students, "Why do you think the table of numbers is called the sieve of Eratosthenes? What happens when we put spaghetti in a sieve?" Students will probably say that the spaghetti stays in the sieve and the water falls through the holes. You can explain that this is what happens in the mathematical sieve—the prime numbers stay in the sieve and the composite numbers fall through the holes (or are crossed out). This activity is an extended version of the technique that Eratosthenes used. Begin the lesson by reviewing the rules of divisibility with students. Using "The Divisibility Ditty" will remind students of the divisibility rules for the numbers 2, 3, 4, 5, 6, 8, 9, and 10.

The Exploration

This activity is done as a whole group activity. The following procedure should be used with students. Walk around the room to make sure that students are able to follow the directions.

Explain to students that they will be finding all of the prime numbers that are less than or equal to 500.

1. Have students cross out the number 1—it is neither prime nor composite. Ask why.
2. Have students highlight the number 2—it is the first prime number. Then have them cross out all the multiples of 2. Ask students, "How will you know if a number is a multiple of 2?" Students should know that the multiples of 2 are all the even numbers, or they could count every second number following the 2. Give students time to complete this procedure. Ask students, "What percentage of the numbers on the chart were eliminated when we removed the multiples of 2? Did you see any particular pattern?"
3. Have students highlight the number 3—it is the next prime number. Now they must cross out all the multiples of 3. Ask students, "Are some of the multiples of 3 already crossed out? How did this happen? Are any of the multiples of 3 still remaining? How can we find these?" Suggest that every sixth number still remaining (following the 3) will be a multiple of 3. Or have students use the divisibility rule: Find the sum of the digits. If the sum is divisible by 2, then the number is divisible by 2. For example, let's examine the number 456—$4 + 5 + 6 = 15$;

15 is divisible by 3 and so 456 is divisible by 3. Give students time to complete this procedure. These two questions can be asked after each step: "Did you see a particular pattern when you eliminated the multiples of three? What method(s) did you use to be sure that you had eliminated the correct numbers?"

4. Ask students what the next prime number is; it's the number 5. Students must highlight this number and cross out all of its multiples. Ask students, "How will you know if a number is divisible by 5? Are any of the multiples of 5 crossed out? Why?" Students should now eliminate all of the multiples of 5. Give students time to complete this procedure.

5. The next prime number is 7. Ask students, "What multiples of 7 have been crossed out? What is the first multiple of 7 that has not been eliminated?" (7 × 7 is the first; 7 × 8, 7 × 9, 7 × 10 are also gone; they were eliminated with the multiple of 2 and 3.) The next multiple of 7 to be eliminated is 7 × 11, then 7 × 13, then 7 × 17, and so forth. Students need only eliminate the multiples of the remaining prime numbers. The last composite number remaining on the chart will be 19 × 23. All the rest of the numbers will be prime!

This is a list of the prime numbers between 1 and 500.

2	3	5	7	11	13	17	19	23	29
31	37	41	43	47	53	59	61	67	71
73	79	83	89	97	101	103	107	109	113
127	131	137	139	149	151	157	163	167	173
179	181	191	193	197	199	211	223	227	229
233	239	241	251	257	263	269	271	277	281
283	293	307	311	313	317	331	337	347	349
353	359	367	373	379	383	389	397	401	409
419	421	431	433	439	443	449	457	461	463
467	479	487	491	499					

Debriefing

Discuss with students how they can be sure that all of the prime numbers between 1 and 500 are now highlighted on their chart. Ask them if they see any pattern.

ASSESSMENT

1. Student products: Assess students' work on the prime number chart.
2. Journal questions:
   a. Level 1 question: Explain, in your own words, the process we used to find all of the prime numbers between 1 and 500.
   b. Level 2 question: If the chart had 1000 numbers on it, what is the last composite number you would need to remove before the remainder of the numbers were all prime? Explain how you solved this problem.

500 Chart

Directions: Use this chart to find all of the prime numbers ≤ 500. Highlight the prime numbers and cross out all of the multiples of that number. Only primes will remain.

1	2	3	4	5	6	7	8	9	10	11	12	13	14	15	16	17	18	19	20	21	22	23	24	25
26	27	28	29	30	31	32	33	34	35	36	37	38	39	40	41	42	43	44	45	46	47	48	49	50
51	52	53	54	55	56	57	58	59	60	61	62	63	64	65	66	67	68	69	70	71	72	73	74	75
76	77	78	79	80	81	82	83	84	85	86	87	88	89	90	91	92	93	94	95	96	97	98	99	100
101	102	103	104	105	106	107	108	109	110	111	112	113	114	115	116	117	118	119	120	121	122	123	124	125
126	127	128	129	130	131	132	133	134	135	136	137	138	139	140	141	142	143	144	145	146	147	148	149	150
151	152	153	154	155	156	157	158	159	160	161	162	163	164	165	166	167	168	169	170	171	172	173	174	175
176	177	178	179	180	181	182	183	184	185	186	187	188	189	190	191	192	193	194	195	196	197	198	199	200
201	202	203	204	205	206	207	208	209	210	211	212	213	214	215	216	217	218	219	220	221	222	223	224	225
226	227	228	229	230	231	232	233	234	235	236	237	238	239	240	241	242	243	244	245	246	247	248	249	250
251	252	253	254	255	256	257	258	259	260	261	262	263	264	265	266	267	268	269	270	271	272	273	274	275
276	277	278	279	280	281	282	283	284	285	286	287	288	289	290	291	292	293	294	295	296	297	298	299	300
301	302	303	304	305	306	307	308	309	310	311	312	313	314	315	316	317	318	319	320	321	322	323	324	325
326	327	328	329	330	331	332	333	334	335	336	337	338	339	340	341	342	343	344	345	346	347	348	349	350
351	352	353	354	355	356	357	358	359	360	361	362	363	364	365	366	367	368	369	370	371	372	373	374	375
376	377	378	379	380	381	382	383	384	385	386	387	388	389	390	391	392	393	394	395	396	397	398	399	400
401	402	403	404	405	406	407	408	409	410	411	412	413	414	415	416	417	418	419	420	421	422	423	424	425
426	427	428	429	430	431	432	433	434	435	436	437	438	439	440	441	442	443	444	445	446	447	448	449	450
451	452	453	454	455	456	457	458	459	460	461	462	463	464	465	466	467	468	469	470	471	472	473	474	475
476	477	478	479	480	481	482	483	484	485	486	487	488	489	490	491	492	493	494	495	496	497	498	499	500

Diagramming Divisibility

MATH TOPICS

divisibility rules, multiples, factors, Venn diagrams

PRIOR KNOWLEDGE NEEDED

1. divisibility rules for 2, 3, and 5
2. multiples
3. factors
4. understanding of a three-set Venn diagram

DIFFERENTIATION STRATEGIES

Principles

Flexible grouping: Students work alone to complete the three-set Venn diagram and in pairs to design their own.

Ongoing assessment: Assess successful completion of the three-set Venn diagram and originality of the group's original Venn diagram.

Teacher's Strategies

Product: Tiered Journal Questions: Based upon student readiness, journal question Level 1 or journal question Level 2 can be assigned. Level 1 is easier because it uses a two-set Venn diagram and therefore the least common multiple is easier to find. The Level 2 question is a three-set Venn diagram and so the least common multiple is more difficult to find.

According to Students

Learning Styles/Multiple Intelligences: naturalist, logical/mathematical, interpersonal, intrapersonal

MATERIALS NEEDED

1. an activity sheet for each student
2. one blank three-set Venn diagram for each pair of students

TEACHING SUGGESTIONS

Engaging the Students

Use a three-set Venn model and label one circle "students wearing jeans," the second circle, "students wearing red shirts," and the third circle, "students with brown hair." Ask students to help you place each student in the correct area of the Venn diagram. Discuss whether each of the placements is correct and why. Explain to students that they are going to be using numbers to complete a three-set Venn diagram similar to the one you just did.

The Exploration

Use "The Divisibility Ditty" to review the divisibility rules before students begin the activity. When individuals have completed the activity, collect these sheets and assign partners to design an original three-set Venn problem. They should be permitted to choose from a variety of subjects including plant species, animal species, and so forth.

Debriefing

Allow time for each pair of students to share their original Venn diagram with another group. Time should be allotted for students to discuss their Venn puzzles and any changes that they would like to make in their choices.

This is a correct version of the three-set Venn diagram for the multiples of 2, 3, and 5.

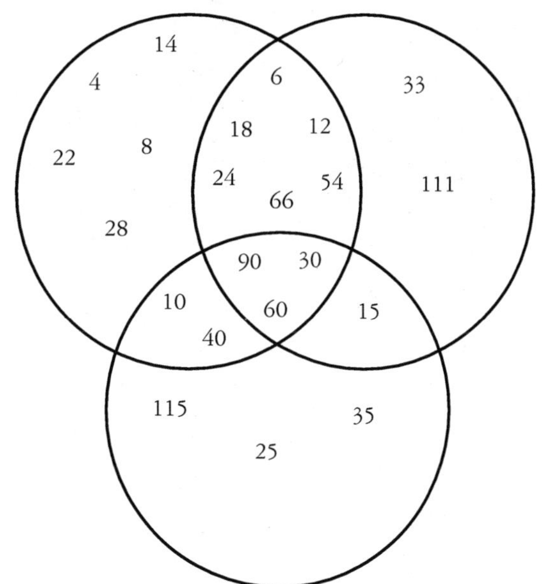

ASSESSMENT

1. Student products: Assess student activity sheets and each group's original three-set Venn problem.
2. Journal questions:
   a. Level 1 question: You are examining a Venn diagram that shows the multiples of 3 in one circle and the multiples of 7 in the other. There is no number in the middle region (where the two circles intersect). Give an example of a number that might be placed correctly in this region. Explain how you got your answer.
   b. Level 2 question: You are examining a Venn diagram that depicts the multiples of 2, 5, and 7. There is no number in the center region (where all three circles intersect). Give an example of a number that might be correctly placed in this region. Explain your answer.

Diagramming Divisibility

Directions: Place the numbers that are at the bottom of the page in the correct circle. If the number has something in common with two circles, place it in the space that is common to both; if it has something in common with all three circles, place it in the space that indicates this. If the number has nothing in common with any of the factors, place it outside the circles.

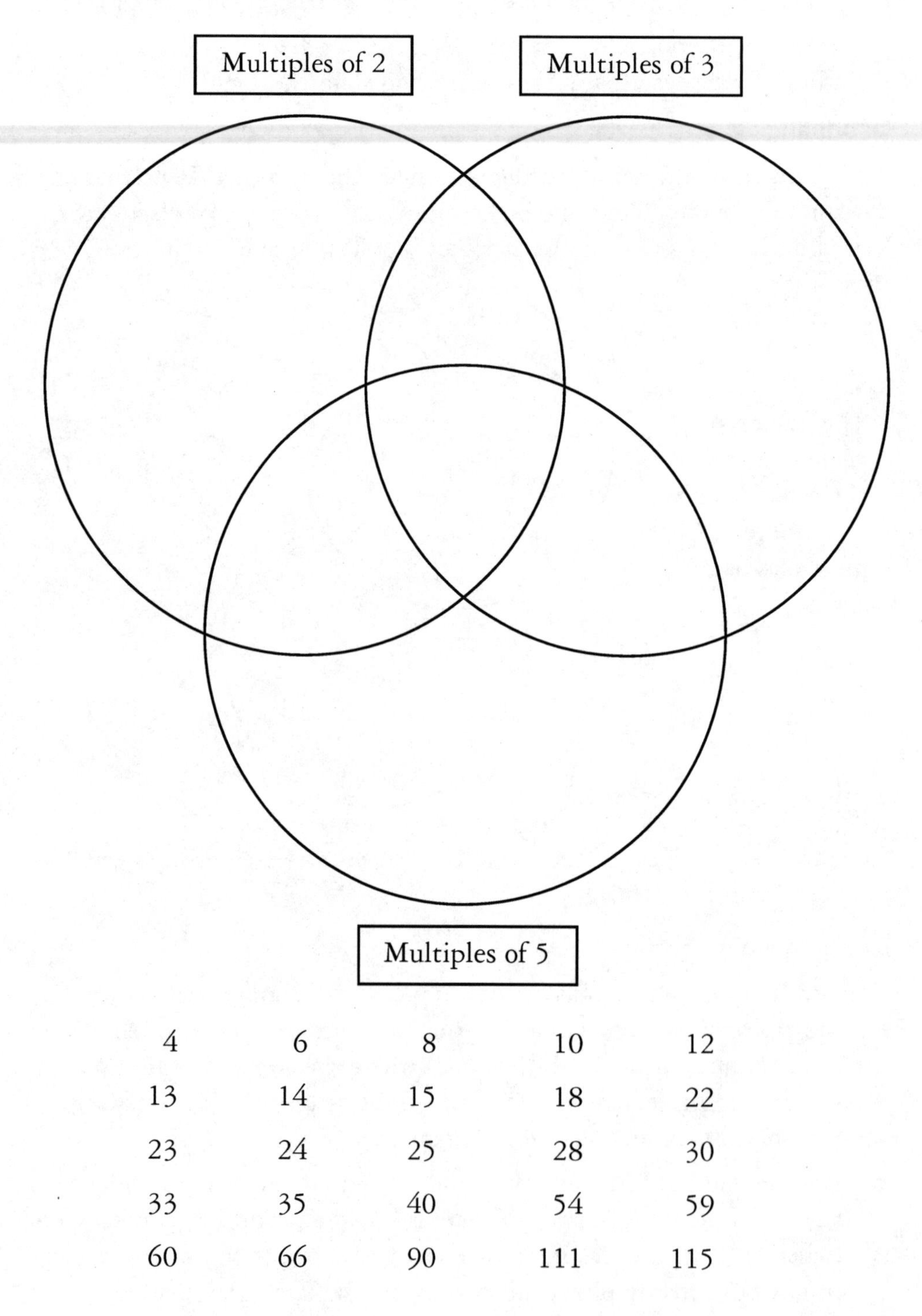

4	6	8	10	12
13	14	15	18	22
23	24	25	28	30
33	35	40	54	59
60	66	90	111	115

Diagramming Divisibility

Directions: Work with your partner and use the Venn diagram below to design your own Diagramming Divisibility activity sheet. Be sure to create an answer key so when you share it with the class you can help with the solutions.

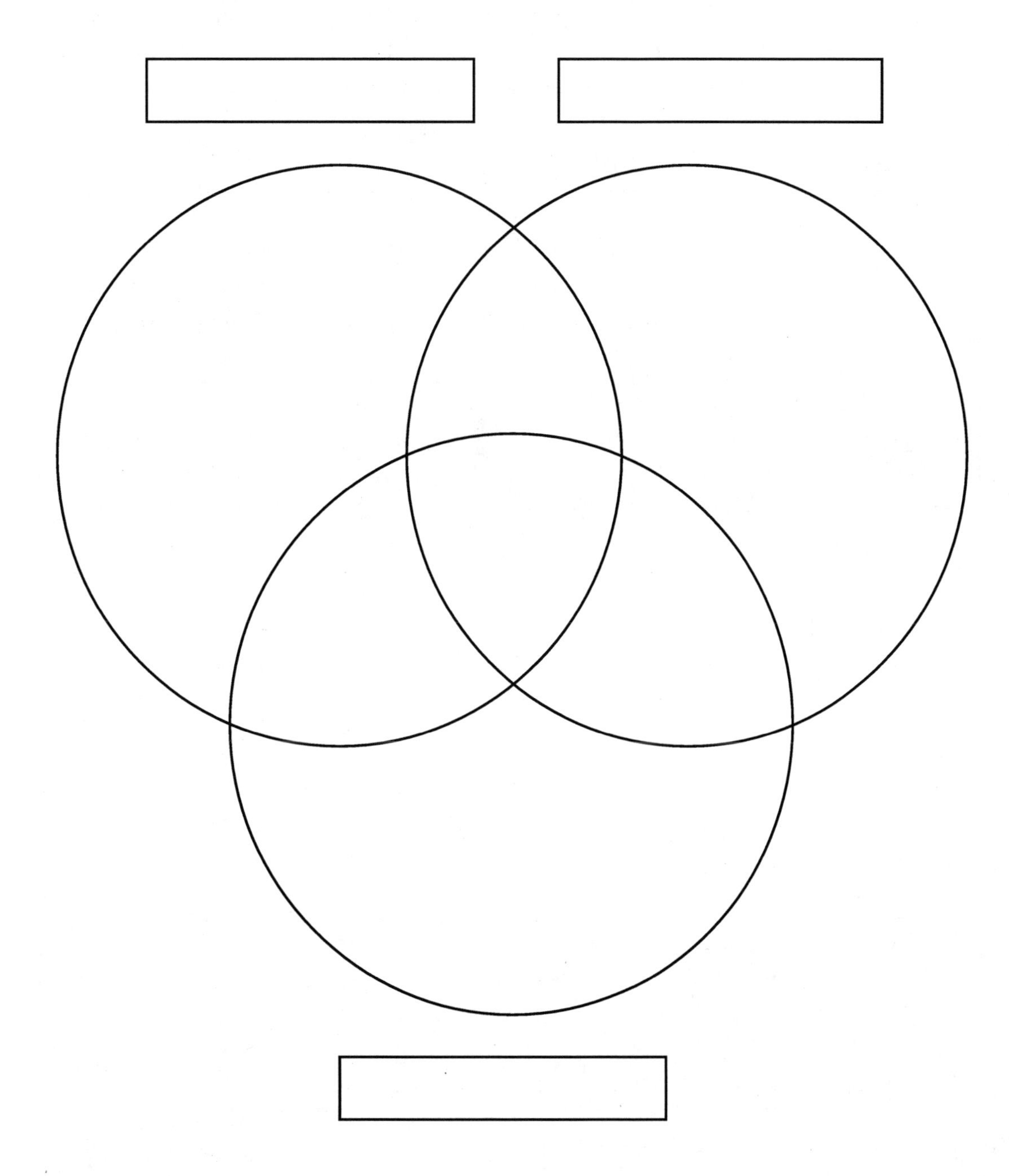

Brush Up Those Skills
Chapter 1

Directions: Roll one die; your group will work on the activity that matches the number on the die. If you have already completed that activity, roll the die until it lands on an activity you have not yet done. On the next page, check off the activity your group completed. Make sure each group member has signed his or her name.

1	Find the year that your state joined the Union. Use the four digits of the year and order of operations to find the numbers 1 through 10. You may use addition, subtraction, multiplication, division, parentheses, exponents, or square roots. You can change the order of the numbers but you must use each number only once.
2	Ask your teacher for a copy of Pascal's triangle. Each member of your group is to choose from one of these numbers: 2, 3, 4, 5, or 6. Using colored pencils or markers, shade in all of the multiples of your number. Now compare the designs. Write a description of each of the designs—how are they alike and how are they different?
3	What's the big deal about order of operations? Does it really make a difference? To answer this question, let's examine these two problems: A. $(4 + 5) \times 6$ and B. $4 + (5 \times 6)$. When we solve these, do we get the same answer? Work with your group to write a story problem for each of these problems.
4	Did you know that the land snail is the slowest animal on Earth? It travels only 8 inches in 1 minute. What is the snail's rate in miles per hour? Work with your group to solve this problem. Be sure to explain how you converted from inches per minute to miles per hour.
5	In an acrostic poem, the letters of the title are written vertically, and each line of the poem begins with the corresponding letter of the title. The poem must refer to a word; it cannot merely be a collection of letters. Work with your group to design an acrostic poem that refers to factors, multiples, prime numbers, or another number concept.
6	A recipe for cookies calls for $2\frac{1}{2}$ cups of flour, $\frac{3}{4}$ cup of sugar, $1\frac{1}{4}$ cups of brown sugar, 2 cups of chocolate chips, 1 cup of margarine, 1 egg, 1 teaspoon vanilla, 1 teaspoon baking powder, and 1 teaspoon baking soda. This recipe makes about 4 dozen cookies. Rewrite this recipe so that it will make 6 dozen cookies.

Brush Up Those Skills Chapter 1

Our Progress Completing These Activities

We have completed | **Signatures of Group Members**

☐ Activity 1 ______________________
____
____
____

☐ Activity 2 ______________________
____
____
____

☐ Activity 3 ______________________
____
____
____

☐ Activity 4 ______________________
____
____
____

☐ Activity 5 ______________________
____
____
____

☐ Activity 6 ______________________
____
____
____

Patterns, Functions, and Algebra

In mathematics he was greater
Than Tycho Brahe, or Erra Pater:
For he, by geometric scale,
Could take the size of pots of ale;
Resolve, by sines and tangents straight,
If bread or butter wanted weight
And wisely tell what hour o' th' day
The clock does strike by algebra.

—SAMUEL BUTLER (1612–1680)

When students study algebra the most common question usually is, "When are we ever going to have to use this?" Rather than present algebra as a set of rules and procedures, the lessons in this chapter relate algebra to real-world phenomena and motivating activities.

We start off with a contest—which pair of students can land their parachutes the closest to each other? In the "The Parachute Jump," students work in pairs to design parachutes and use the Pythagorean theorem to calculate the distance between their "jumps." Using the song "Ode to Pythagoras" (page 37) and their active involvement, students begin to see the real-world applications of algebra while using a variety of learning styles to better understanding this concept.

The Pythagorean theorem is revisited in "The Irrational Spiral" (page 43), but this time students experiment with algebra, geometry, and design. The lesson begins by "replaying" the musical ditty "Ode to Pythagoras." Students, using protractors and rulers, draw a right triangle with a base and height of 1 cm and progress to design an amazing spiral design.

The activity "Seeing to the Horizon" (page 44) explains the scientific principle that on a clear day, "the higher up you are, the further you can see." A list of the ten tallest skyscrapers is provided and, using an algebra formula, students calculate how far they can see from the rooftops of these buildings. This is the perfect time to work with the art teacher and introduce perspective and vanishing points to students.

Students continue to research the height of their school building, but this time using stopwatches and marshmallows. In "Falling Objects: Formulas in the Real World" (page 49), students start out with skyscrapers and then use marshmallows, stopwatches, and algebraic formulas to estimate the height of their school in a unique way.

And finally, "Pattern Block Patterns" (page 55) uses hands-on manipulatives to have students build geometric patterns. By describing, analyzing, and replicating the pattern, they are using algebraic reasoning while utilizing different learning styles and intelligences.

Cokie Roberts is quoted as saying, "As long as algebra is taught in school, there will be prayer in school." Perhaps if students are given the opportunity to experience hands-on activities related to algebra, the subject will not generate such aversion and apprehension.

The Parachute Jump

MATH TOPICS

ordered pairs on the coordinate plane, the Pythagorean theorem, data collection and analysis

PRIOR KNOWLEDGE NEEDED

1. understanding of (x, y) coordinates on the coordinate plane
2. knowledge of the Pythagorean theorem

DIFFERENTIATION STRATEGIES

Principles

Flexible grouping: For this activity, students work in pairs. While this activity is designed to help students calculate the distance between two points on the coordinate plane using the Pythagorean theorem, the activity can be tiered. More advanced students can be taught the distance formula and use the formula to find the distance. $D = \sqrt{(x_2 - x_1)^2 + (y_2 - y_1)^2}$.

Ongoing assessment: Observe students during the activity, and use data collection sheets to assess level of understanding.

Teacher's Strategies

Product: Tiered Journal Questions: The Level 1 question is an easier question because it asks students to explain, in their own words, what happened in the experiment that uses the Pythagorean theorem. The Level 2 question is written for students who used the distance formula. It asks them to explain why the distance formula is a shortened version of the Pythagorean theorem.

According to Students

Learning Styles/Multiple Intelligences: logical/mathematical, bodily/kinesthetic, musical/rhythmic, visual/spatial, verbal/linguistic, interpersonal

MATERIALS NEEDED

Each pair of students will need the following:

1. two coffee filters

2. eight 12-inch lengths of string
3. a paper punch
4. large safety pins, large paper clips, or washers
5. colored pencils or markers
6. poster board or a large sheet of graph paper
7. a student activity sheet
8. data collection table

TEACHING SUGGESTIONS

Engaging the Students

Discuss with students the parachute jump experiment. Have two parachutes made to represent the two students' parachutes. Drop one previously made parachute onto a coordinate plane and mark that point. Now drop the second one; mark that point as well. Lightly draw a line connecting the two drop points. Discuss with students that the length and width of each of the squares on the graph represent a side of 1 and ask, "Can the distance between the two drop points be measured as if it were the side of a square?" Using a pencil, lightly draw a vertical line from one drop point and a vertical line from the other drop point. These two lines will form the legs of a right triangle—the line connecting the distance between the two drop points is the hypotenuse of this triangle. Discuss with students how the Pythagorean theorem can be used to approximate the length of the distance between the drop points.

The Exploration

The song "Ode to Pythagoras" can be sung before this experiment is conducted. This is an interesting way to review the Pythagorean theorem and is most effective with students with a variety of learning styles.

Students are placed into pairs and given one student activity sheet. This sheet contains a set of directions to help them design and build their own parachute to use in the experiment and fill out the data collection table. Four holes are evenly spaced around the edge of a coffee filter, and string (about 12 inches long) is attached to each hole. A large paper clip or washer must be attached to the string to help the parachute land. Students may decorate their parachutes for purposes of identification.

Each team must divide a poster board into 4 quadrants, labeling the x and y coordinate points unless large sheets of graph paper are available for their use. (An overhead transparency can serve as a sample for the groups.)

Each team member, taking turns, drops his or her parachute onto the poster board. Each landing location (the x and y coordinate point) is recorded in the data collection table.

It is possible to form a right triangle by connecting the two landing points (this line becomes the hypotenuse of the right triangle) and then drawing the horizontal and vertical lines from these points to form the legs of the triangle. The lengths of the vertical line (side b) and horizontal line (side a) can be easily calculated and the Pythagorean theorem ($a^2 + b^2 = c^2$) can be used to find the distance between the two landing points. For example, if the two parachutes land on the coordinates indicated on this graph, the length of side $a = 2$ and the length of side $b = 3$; $2^2 + 3^2 = c^2$. $4 + 9 = c^2$; $c = \sqrt{13} \approx 3.6$. For those students using the distance formula, this formula can be easily substituted in the data collection table in the last column. Students will conduct ten drops and find the average or mean distance between them.

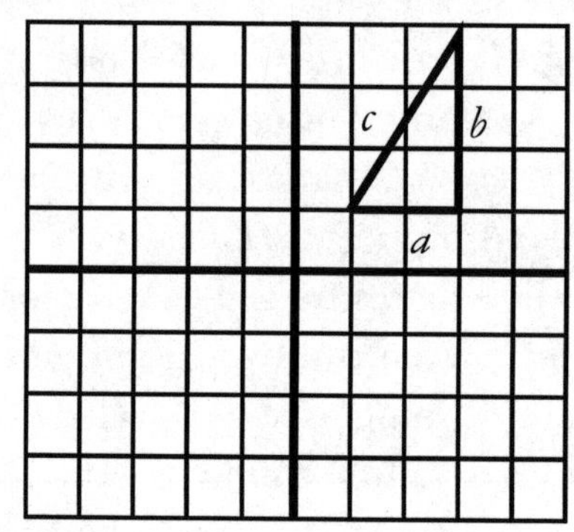

Debriefing

Give each group time to discuss their results and the mean distance between the ten drops. You can award a prize to the team with the shortest mean distance.

ASSESSMENT

1. Student products: Make sure students have successfully completed activity sheets.
2. Journal questions:
   a. Level 1 question: Discuss the interesting aspects of your experiment by describing 1) how you used the Pythagorean theorem to help you find the distances, 2) why you could not just measure the distance between the two coordinate points, 3) your results (if landings in one quadrant occurred more often than another, if the parachute drifted or always landed in the same location, and why you think this occurred).
   b. Level 2 question: Analyze both the Pythagorean theorem and the distance formula. Explain why the distance formula is actually the Pythagorean theorem expressed in another way and why either one can be used to find the hypotenuse of a right triangle.

Ode to Pythagoras

SUNG TO "HOKEY POKEY"

You take your first leg "*a*,"
And then your next leg "*b*,"
Take the sum of their squares,
Are you following me?

To use this famous theorem,
For Pythagoras let's shout,
That's what it's all about.

We're not quite finished yet,
Cause there's a third side to see,
You know this tri-angles right,
Are you following me?

To use this famous theorem,
For Pythagoras let's shout,
That's what it's all about.

Now add the square of leg "*a*,"
To the square of leg "*b*,"
You get the square of side "*c*,"
Are you following me?

To use this famous theorem,
For Pythagoras let's shout,
That's what it's all about.

$$a^2 + b^2 = c^2$$

The Parachute Jump

Directions: Work with a partner to complete the following activity.

A. You will need:

1. one coffee filter each
2. four 12-inch lengths of string each
3. a large paper clip, a large safety pin, or a washer each
4. one piece of poster board or large sheet of graph paper for you and your partner
5. one data collection table for you and your partner
6. colored markers or colored pencils

B. You will need to create and design your own parachute. Punch four holes near the edge of a coffee filter at evenly spaced intervals. Attach four pieces of string (12 inches long) in each of the holes. Draw the four pieces of string together and attach a large paper clip, a safety pin, or a washer at the end of the connected string. Be sure to decorate your parachute for ease in identification.

C. Work with your partner to create a "landing area" for the parachutes.

1. Divide the poster board or your large sheet of graph paper into 4 quadrants.
2. Label the x and y coordinate points.

D. Drop your parachutes (one at a time) from the designated height above the landing area. The "landing location" is the nearest (x, y) point on the "landing area." Record ten successive drops for you and your partner on the data collection table.

E. Use graph paper after each "drop" to plot the coordinate points for your drop and your partner's drop. Connect the two points, and draw a horizontal line from one of the points and a vertical line from the other. You have now formed a right triangle. Use the Pythagorean theorem to find the length of the hypotenuse of this triangle.

F. After your ten trials, find the mean difference of distance between the parachute drops. Compare your results with other groups. How did your distances compare?

The Parachute Jump

DATA COLLECTION

Trial Number	(x, y) Coordinates of Team Member 1	(x, y) Coordinates of Team Member 2	Length of Side a	Length of Side b	Distance Between Landing Locations of Parachutes 1 and 2: $a^2 + b^2 = c^2$
1					
2					
3					
4					
5					
6					
7					
8					
9					
10					
Mean					

The Irrational Spiral

MATH TOPICS

Pythagorean theorem, substitution for a variable, angle measurement, irrational numbers, problem solving

PRIOR KNOWLEDGE NEEDED

1. knowledge of the Pythagorean theorem
2. knowledge of use of protractor
3. understanding of rational and irrational numbers

DIFFERENTIATION STRATEGIES

Principles

Flexible grouping: Students work in pairs to draw the irrational spiral.

Ongoing assessment: Observe student groups and ask productive questions such as "Can you explain the steps you are using to draw each of these right triangles?" or "How are each of these triangles the same? How are they different?" or "Why is it important to measure each of the right angles rather than draw them freehand?"

Teacher's Strategies

Product: Tiered Journal Questions: The Level 1 question asks students to explain the difference between rational and irrational numbers and to give an example of each. The Level 2 question is more difficult because it requires students to convert the repeating decimal $0.\overline{142857}$ to a fraction and explain why the irrational number π cannot be expressed in the same form.

According to Students

Learning Styles/Multiple Intelligences: visual/spatial, logical/mathematical, musical/rhythmic, interpersonal, bodily/kinesthetic

MATERIALS NEEDED

1. large sheet of construction paper for each pair of students
2. protractors

3. metric rulers
4. colored pencils

TEACHING SUGGESTIONS

Engaging the Students

There are many spirals that occur in nature. Two of these are the spiral formed by Fibonacci numbers (found on pinecones and pineapples) and the spiral that appears in the nautilus shell. An example of each of these can be found on the Internet for student exploration.

The Exploration

Introduce the Pythagorean theorem with the song, "Ode to Pythagoras." Review the theorem, and be sure that all of the students understand how to solve the equation and use square roots.

Have the students describe the irrational spiral in their journals and then share their descriptions with the class.

Students are asked to find $\sqrt{3}$. If available, use an overhead calculator to discuss if the number appears to be a terminating or a repeating decimal. Explain to students that this type of number is called *irrational* and does not terminate or repeat. Have students round this number to the nearest hundredth. Go around the spiral, and convert each of the irrational numbers to rational approximations.

Allow students to work in pairs and have them draw an irrational spiral as large as possible on the paper. Tell them that it will start to wrap around and they will not be able to continue connecting their lines at the center. Have them stop at the edge of the design so that the lines do not overlap.

It is essential that students' measurements be as accurate as possible or the spiral will not look correct. Students should use the Pythagorean theorem to find the new side and write down that proof—it should be turned in with the drawing for full credit.

Debriefing

Give students an opportunity to display their spirals. Discuss the reasons for using radical expressions rather than approximations.

ASSESSMENT

1. Student products: Drawings of irrational spirals can be used as part of the assessment process.
2. Journal questions:
   a. Level 1 question: Explain the difference between rational and irrational numbers. Be as specific as you can and give an example of each. Be sure to explain why each of these numbers is rational or irrational.
   b. Level 2 question: $0.\overline{142857}$ is a rational number because it can be expressed in the form of *a/b*. 1) Express this number as a fraction, and 2) explain why the irrational number π cannot be expressed in the same form.

VARIATIONS FOR DIFFERENTIATION

It is possible to cut out each triangle of fabric and sew an irrational spiral quilt. Use 1-cm graph paper as a pattern. Give each triangle an additional $\frac{1}{4}$ inch for seam allowance. It can be made about the same size as the drawing on the student activity sheet.

The Irrational Spiral

The Irrational Spiral is formed by careful measurement using a protractor and a metric ruler. Start with a right triangle with sides of 1 cm. When you draw the hypotenuse, you have a length of $\sqrt{2}$. Let's see why this happens.

Using the Pythagorean theorem, we know that $a^2 + b^2 = c^2$. Since each side is equal to 1 by substitution, we see that $1^2 + 1^2 = c^2$. So $c^2 = 2^2$ and $c = \sqrt{2}$. By using the hypotenuse of this triangle as one of the sides of a new triangle (the other leg remains 1), the length of the new hypotenuse is $1^2 + \sqrt{2}^2 = 3^2$. $1 + 2 = 3^2$. So $c = \sqrt{3}$. An example of this irrational spiral is shown below.

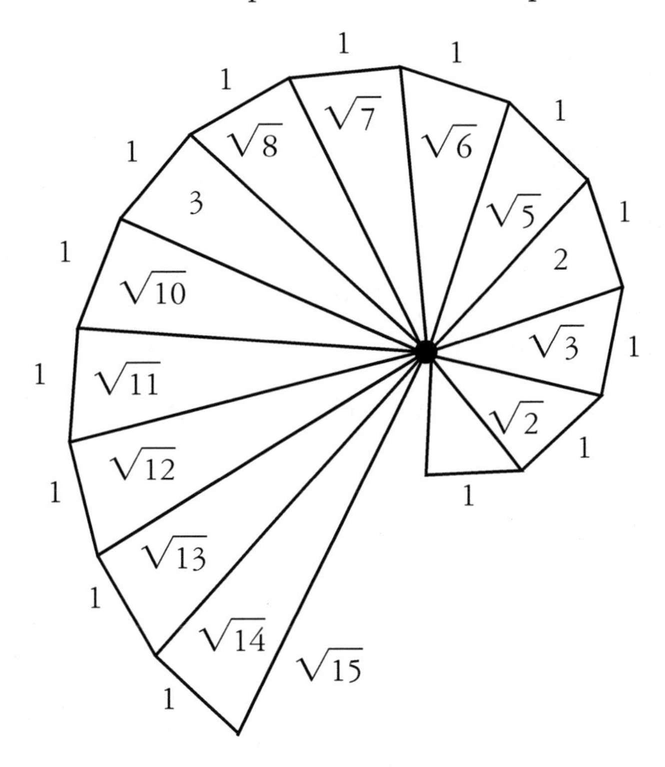

Use your calculator to find $\sqrt{3}$ or $\sqrt{5}$. What numbers appear on the display? Does the decimal appear to terminate or repeat? Can you see any pattern to the numbers? Because these decimals do not terminate and do not have a repeating pattern, the sides that do not have a whole number (rational root) are called irrational numbers. What irrational numbers are pictured above?

Some of the sides shown are rational numbers. What will the next rational number length be?

Use a large sheet of paper and see how large an irrational spiral you can design!

Seeing to the Horizon

MATH TOPICS

using formulas, computation, square roots, reading data from a table, averages, problem solving

PRIOR KNOWLEDGE NEEDED

1. substituting for a variable in a formula
2. using a calculator to find the square root of a number
3. finding the mean

DIFFERENTIATION STRATEGIES

Principles

Flexible grouping: Students will work individually to complete the activity sheet and then work as a group of four to estimate the height of the school.

Ongoing assessment: Observe students, and use productive questioning to focus their attention. For example, "What prior knowledge or information about the school might you use to help you estimate the building's height?" or "How many feet do you think the average story in a building might be?"

Teacher's Strategies

Product: Tiered Journal Questions: Use the Level 1 and Level 2 journal questions to meet the needs of students at various readiness levels. The Level 2 question is multistep, whereas the Level 1 question is very similar to those in this activity.

According to Students

Learning Styles/Multiple Intelligences: logical/mathematical, verbal/linguistic, interpersonal

MATERIALS NEEDED

1. one student activity sheet for each student
2. calculators

3. an overhead transparency of class data table
4. tape measures, rulers, or yardsticks for estimation

TEACHING SUGGESTIONS

Engaging the Students

Explain to students that this activity has two separate parts. The first asks students to imagine that they are on the top of the ten tallest skyscrapers in the world and are looking toward the horizon. Ask, "Where in the world do you think these skyscrapers will be?" They might be surprised to learn that eight out of the ten are in either the Near East or the Far East—currently only the Sears Tower in Chicago and the Empire State Building in New York are in the United States.

The Exploration

Give students the activity sheet "Seeing to the Horizon" and go over the formula at the top of the page. Explain that 1.22 is a constant and is based upon physical principles relating to the earth and its environment. Then ask students to complete the data table using calculators. Review the results as a group. The answers are in the table below (rounded to the nearest mile).

10 of the World's Tallest Buildings

Building	Height (ft)	Distance You Can See to the Horizon (mi)
Taipei 101	1671	50
Petronas Towers 1 & 2	1483	47
Sears Tower	1450	46
Jin Mao Tower	1380	45
Two International Finance	1362	45
CITIC Plaza	1283	44
Shun Hing Square	1260	43
Empire State Building	1250	43
Central Plaza	1227	43
Bank of China Plaza	1205	42

Source: emporis.com/en/bu/sk/st/tp/wo/

Once students understand how to use the formula and have had some practice using it, they can be placed into groups. Groups will estimate the height of their own school building. If there is a more appropriate building (or structure) for them to use, this will serve the same purpose.

Working in groups of four, using tape measures, yardsticks, and so forth, students will estimate the height of the school or other structure. Some groups may need to be guided through questioning to get them on the right track. Their estimates should be recorded on the class data collection table and discussed during debriefing.

Debriefing

Once each group has entered their estimations on the class data collection table, ask students to discuss the data and to look for any problems that they might see. Do all of the answers make sense? How did students decide how tall the building is? What is the range of the estimates? Do some estimates appear to be outliers (far from the mean)? Before an average or mean is found, should some of the estimates be eliminated? Once all of the data is accepted and a class mean is found, this estimate can be used to calculate the distance to the horizon from the top of the school. Round to the nearest $\frac{1}{10}$ mile.

ASSESSMENT

1. Student products: Student activity sheets and group estimation sheets can be used as part of the assessment process.
2. Journal questions:
    a. Level 1 question: Mountains are generally measured from sea level, in which case Mount Everest is the tallest measuring 29,028 feet. On a clear day, how far could you see to the horizon from the top of Mount Everest?
    b. Level 2 question: Two famous volcanoes are Mauna Kea in Hawaii and Mount St. Helens in Washington. Mauna Kea is 29,568 feet above sea level and Mount St. Helens is 8364 feet above sea level. If you stood on the top of each of these volcanoes, how much farther to the horizon could you see from Mauna Kea than from Mount St. Helens?

Seeing to the Horizon

The distance you can see to the horizon on a clear day is given by the formula

$$d = 1.22\sqrt{h}.$$

In this formula, d represents the distance in miles and h represents the height in feet your eyes are from the ground. In other words, the higher you are, the farther you can see.

Directions: Use this formula to calculate the distance you can see to the horizon if you are standing on the roof of one of these buildings.

10 of the World's Tallest Buildings

Building	Height (ft)	Distance You Can See to the Horizon (mi)
Taipei 101	1671	
Petronas Towers 1 & 2	1483	
Sears Tower	1450	
Jin Mao Tower	1380	
Two International Finance	1362	
CITIC Plaza	1283	
Shun Hing Square	1260	
Empire State Building	1250	
Central Plaza	1227	
Bank of China Plaza	1205	

Source: emporis.com/en/bu/sk/st/tp/wo/

Now, work with your group to estimate the height of your school building.

Write your estimate (in feet) here: ______________________________

Write your estimate on the class data collection table so we can find the class mean.

Seeing to the Horizon

CLASS DATA

Group	Estimate of the Height of the Building (ft)	How Far Can We See to the Horizon? (ft)
Class Mean		

Be prepared to discuss the steps your group used to problem-solve the height of your school building.

Falling Objects: Formulas in the Real World

substitution for a variable, using formulas, square roots, rounding, problem solving, conversions

PRIOR KNOWLEDGE NEEDED

1. substituting for variables and using formulas
2. using square roots
3. rounding
4. converting from feet/second to miles/hour
5. ability to read a stopwatch

DIFFERENTIATION STRATEGIES

Principles

Flexible grouping: Students complete the activity sheet individually. When students all understand how to make substitutions into the formula, they are placed into groups of four to conduct the experiment.

Ongoing assessment: Observe students and use productive questioning.

Teacher's Strategies

Product: Tiered Journal Questions: Journal question Level 1 asks students to work with large numbers and convert seconds to days. It is meant to help students practice conversion within the English system of measurement. Journal question Level 2 is a multistep problem that asks students to apply the formula and strategies used in the experiment to solve a similar problem.

Product: Tiered Activities: As an extension for more advanced mathematics students, this activity has students 1) research on the Internet, 2) design a data collection table in a spreadsheet program, and 3) calculate (using the computer program) the vertical drop time for roller coasters.

According to Students

Learning Styles/Multiple Intelligences: Logical/mathematical, bodily/kinesthetic, verbal/linguistic, interpersonal

MATERIALS NEEDED

1. a student activity sheet for each student
2. calculators
3. marshmallows
4. stopwatches

TEACHING SUGGESTIONS

Engaging the Students

Before working on this experiment, students will need some background information. The time it takes for something to fall is determined by the formula used to find the vertical drop of an object. It assumes that there is no wind resistance and that the force of gravity is constant on every object (regardless of mass). On Earth, objects accelerate at a rate of 32 feet per second per second. This acceleration is the result of the force of gravity that pulls an object toward the center of Earth. This constant is used in the formula as a substitute for the variable g, but when the fraction is reduced becomes 16.

To find the time it takes to reach the ground, students substitute into the formula:

$$t = \sqrt{\frac{2d}{g}} = \sqrt{\frac{2d}{32}} = \sqrt{\frac{d}{16}}$$

Following the order of operations, students can multiply the height of the building by 2, then divide it by 32 (the acceleration caused by gravity), then find the square root. Or by simplifying the radical, students can divide the height of the building by 16 and then find the square root.

To find the rate of speed of the marshmallow, students must convert ft/sec to mi/hr. One way to do this is to do the following:

1. Find the rate of speed in ft/sec by multiplying the time (column 3) by 32. This rate is ft/sec.
2. Convert ft/sec to mi/hr.

   Computation for the Sears Tower is shown:

   a. 9.5 sec × 32 ft/sec = 304 ft/sec

   b. 304 ft/sec × 3600 sec/1 hr = 1,094,400 ft/1 hr

   c. 1,094,400 ft/1 hr × 1mi/5280 ft ≈ 207 mi/hr

The answers to each of the student activity sheet problems are shown on the table on page 51. Rate of speed has been rounded to the nearest mile per hour.

10 of the World's Tallest Buildings

Building	Height (ft)	Time to Reach the Ground (sec)	Rate of Speed (mi/hr)
Taipei 101	1671	10.2	223
Petronas Towers 1 & 2	1483	9.6	209
Sears Tower	1450	9.5	207
Jin Mao Tower	1380	9.3	203
Two International Finance	1362	9.2	201
CITIC Plaza	1283	9.0	196
Shun Hing Square	1260	8.9	194
Empire State Building	1250	8.8	192
Central Plaza	1227	8.8	192
Bank of China Plaza	1205	8.7	190

Source: emporis.com/en/bu/sk/st/tp/wo/

The Exploration

Students work individually to complete the table. After discussing their results, they are now ready to conduct a real-life experiment working in a group of four. Have students (with proper supervision) or another teacher drop five marshmallows from the roof, one at a time. Using a stopwatch, time how long it takes for the marshmallows to reach the ground to the nearest $\frac{1}{10}$ of a second. Since $d = 16t^2$, we can solve for t by transforming the equation into $t = \sqrt{\frac{d}{16}}$.

Debriefing

When each of the groups has completed the experiment, bring back the students into a whole group setting to discuss their results and explain the data they collected to solve this problem.

ASSESSMENT

1. Student products: Assess the student activity sheet and the group activity sheet for the data collected during the experiment.
2. Journal questions:
   a. Level 1 question: You are asked to write what happened to you 1,000,000 seconds ago. But since seconds is not an appropriate unit of measure for such a large number, you need to convert it into a more suitable unit, perhaps days or months or years. Do the necessary

conversions, making sure you show your work. Then write about what did happen to you 1,000,000 seconds ago.

b. Level 2 question: The Bank of America Plaza, a skyscraper in Dallas, Texas, has 72 floors, each about 13 feet high. Using the formula used in our experiment, find about how long would it take an object dropped from the top of this building to reach the ground.

VARIATION: A TIERED ACTIVITY

Have students research the highest roller coasters in the world. Have students perform an Internet search for roller coaster sites of interest.

Students can design a data collection table in a spreadsheet program—inputting formulas for the vertical drop and rate of speed. After typing in the heights of the coasters, the time and speed will automatically be calculated.

Falling Objects: Formulas in the Real World

When an object falls off a building it is called a "free falling object." To figure out how long it will take to hit the ground, we need to know the distance the object fell. On Earth, objects accelerate at a rate of 32 ft/sec^2. The formula that is used to calculate how long it will take to fall is as follows:

$$t = \sqrt{\frac{2d}{g}} \text{ or } t = \sqrt{\frac{2d}{32}} \text{ or } t = \sqrt{\frac{d}{16}}$$

In the formula, t is the time the objects falls, d is the vertical drop, and g is the acceleration due to gravity. By simplifying the formula, we find that the time it takes for an object to fall is equal to the square root of the distance divided by 16.

Directions: If someone stood on top of each of the buildings listed in the table below and dropped a marshmallow, how long would it take to reach the ground? How fast would the marshmallow be traveling when it hit the ground? Use the vertical drop formula and the data in this table to find the time it would take each marshmallow to reach the ground. Round to the nearest $\frac{1}{10}$ second.

10 of the World's Tallest Buildings

Building	Height (ft)	Time to Reach the Ground (sec)	Rate of Speed (mi/hr)
Taipei 101	1671		
Petronas Towers 1 & 2	1483		
Sears Tower	1450		
Jin Mao Tower	1380		
Two International Finance	1362		
CITIC Plaza	1283		
Shun Hing Square	1260		
Empire State Building	1250		
Central Plaza	1227		
Bank of China Plaza	1205		

Source: emporis.com/en/bu/sk/st/tp/wo/

Falling Objects: Formulas in the Real World

DATA COLLECTION

Now let's conduct a little experiment. How tall do you think your school building is? Discuss this with your group, and write your estimate here:

____

1. Five marshmallows will be dropped from the roof of your school. Using your stopwatches, time their fall as carefully as you can—to the nearest $\frac{1}{10}$ of a second.
2. Record the five trials on this data collection table.
3. Find the average.
4. Use a variation of the vertical drop formula ($d = 16t^2$) to calculate the distance the marshmallow traveled. Use your group average for t.

Trial	Time (sec)
1	
2	
3	
4	
5	
Average	

Based upon our data, we calculate the height of the school to be ___________
Our work:

Pattern Block Patterns

MATH TOPICS

patterns, sequences, algebraic formulas

PRIOR KNOWLEDGE NEEDED

experiences translating concrete patterns to abstract symbols (finding the *n*th term)

DIFFERENTIATION STRATEGIES

Principles

Flexible grouping: Each student works with a partner to find these patterns.

Ongoing assessment: Observe students and use productive questioning.

Teacher's Strategies

Product: Using Manipulatives in a Journal Question: Visual and tactile learners are given the opportunity to design an original problem and then solve it using the same manipulatives used in the activity.

According to Students

Learning Styles/Multiple Intelligences: logical/mathematical, visual/spatial, verbal/linguistic, bodily/kinesthetic, interpersonal, and naturalist

MATERIALS NEEDED

1. a collection of pattern blocks for each pair of students
2. a student activity sheet for each pair of students
3. an overhead transparency of student activity sheet

TEACHING SUGGESTIONS

Engaging the Students

Ask students to build each of the four terms shown. If overhead pattern block tiles are available, demonstrate the problem for the class. Different combinations of pattern blocks can be used to illustrate how these geometric series develop. Ask students, "Do you think you would want to use the same process we have used to solve these first few problems to find the 100th pattern? Why or why not?"

The Exploration

As students start to work in pairs, some productive questions to help students keep on track are 1) What will the next pattern look like? 2) How many tiles will you need to construct it? 3) How many triangles will there be? 4) How many trapezoids? Questions like these help students focus on the numerical pattern and will help them determine the 20th of the series, the 100th of the series, and finally the *n*th term.

Debriefing

Discuss with students the strategies they used to find the 20th or the 100th or the *n*th term of the series.

ASSESSMENT

1. Student products: The student activity sheet can be used for assessment.
2. Journal question: Using two or more of the pattern blocks, develop your own geometric pattern series. Describe 1) how many of each block you will need for each term of your series, 2) how the number of blocks increases each time, 3) the number of blocks in the 20th term, and 4) a formula that you can use to find the *n*th term of your series."

Pattern Block Patterns

Directions: Fill in the lines below to show the number of trapezoids and triangles in each term.

Term 1	Term 2	Term 3	Term 4
__ triangles __ trapezoids	__ triangles __ trapezoids	__ triangles __ trapezoids	__ triangles __ trapezoids

Now, work with your partner to construct each of these geometric patterns. Draw what the next pattern will look like:

Term 5
___ triangles ___ trapezoids

Now, visualize what the 20th geometric pattern will look like. How many triangles will there be? _____ How many trapezoids? _____

Suppose there were 100 terms. How many tiles would there be? _____
How many of them would be triangles? _____ trapezoids? _____

Work with your partner to determine a formula that you could use to find any number of terms—even the *n*th term of the series. Our formula is this:

____

Brush Up Those Skills Chapter 2

Directions: Roll one die; your group will work on the activity that matches the number on the die. If you have completed that activity, roll the die until it lands on an activity you have not done. On the next page, check off the activity your group completed and make sure each group member has signed his or her name.

1	1. Choose any 3 × 3 square on a calendar. Work with your group to discover a way to find the sum of all of the numbers without adding them in the traditional way. 2. If you know the sum of a 2 × 2 square on a calendar is 40, work with your group to find an algebraic method to find the numbers.
2	Choose a number. Multiply by 3. Add 6. Multiply by 2. Divide by 6. Subtract the number you started with. The number you are left with is 2. Now work with your group to make up your own problem using algebra magic!
3	Examine these two patterns: 2, 4, 6, 8, . . . and 2, 4, 8, 16, . . . 1. Give the next five numbers in each pattern. 2. Explain how these patterns are alike and how they are different. 3. Make up two number sequences. Explain how they are alike and how they are different.
4	1. Consider this scale: Find possible values for each of the shapes so that the scale remains balanced. 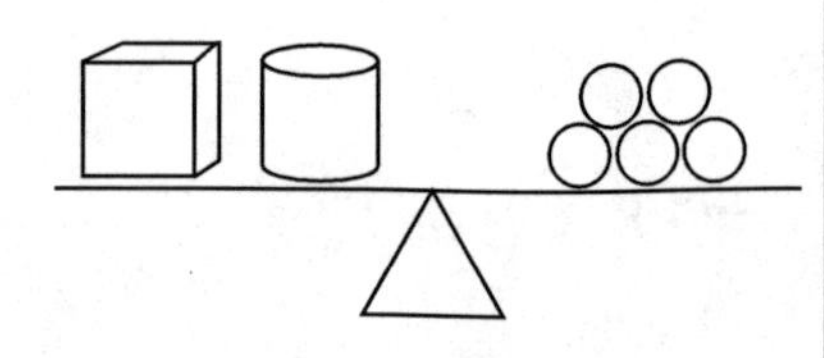 2. Design your own scale, and share it with another group.
5	1. You have 3 boxes of cookies and 6 extra cookies. Altogether you have 42 cookies. How many cookies are in each box? Set up an algebraic equation to help you solve this problem. 2. Make a story problem for this algebraic equation: $5x - 2 = 18$.
6	1 3 6 10 15 These are the first five triangular numbers. How many dots will there be in the next triangular number? 1. Make a table and find the next four triangular numbers. 2. Then find an equation that will help you find the 50th triangular number.

Brush Up Those Skills Chapter 2

Our Progress Completing These Activities

We have completed | Signatures of Group Members

☐ Activity 1 ______________________

____

____

____

☐ Activity 2 ______________________

____

____

____

☐ Activity 3 ______________________

____

____

____

☐ Activity 4 ______________________

____

____

____

☐ Activity 5 ______________________

____

____

____

☐ Activity 6 ______________________

____

____

____

Measurement and Geometry

[The universe] cannot be read until we have learnt the language and become familiar with the characters in which it is written. It is written in mathematical language, and the letters are triangles, circles and other geometrical figures, without which means it is humanly impossible to comprehend a single word.

—GALILEO GALILEI (1564–1642)

Measurement and geometry are the strands of mathematics that help people describe the physical world in which they live. The study of geometry helps students see the world of mathematics in a different light—the mathematics of shapes and forms, not numbers and formulas.

The study of measurement requires students to directly interact with their environment. When students learn to measure, they can answer questions such as, "How much will that hold?" or "How tall is that?" They learn about weight, distance, and time. They develop measurement sense as they acquire a true understanding of space and size. The activities in this chapter allow students to be actively involved, to work collaboratively, and learn more about their physical world as they develop their multiple intelligences.

"Pennies and the Sears Tower" (page 62) encourages students to design an experiment to help them consider the height of the Sears Tower in Chicago in a most unusual way—its value in pennies! As they design and carry out their experiments, they learn, firsthand, about open-ended problems and problems that have more than one correct answer.

Does the height from which a ball is dropped affect how high it bounces? Students answer this intriguing question by conducting the activity "The Bouncing Ball" (page 67). Working collaboratively, students collect, organize, analyze, and graph their data, and in the process, learn a variety of measurement skills and concepts.

"Soda Pop Math" (page 72) combines the study of two- and three-dimensional geometry. Students start the project with a flat sheet of poster board and end with a product can that conforms to consumer needs and nutritional values. This is a truly interdisciplinary activity that supports a variety of intelligences while giving students hands-on experiences with the attributes of a polyhedron.

Paper folding is a remarkable way to demonstrate the attributes of geometric shapes using hands-on, informal strategies. Students can feel and experience the shapes and sizes of different geometric shapes as they fold and manipulate paper into polygons. The activity "Paper-Folding Polygons" (page 77) gives students the opportunity to experience regular pentagons, hexagons, and octagons and then use the concepts of symmetry to transform the shape into a not-so-ordinary snowflake.

While mathematics is often taught as separate disciplines, there is an interesting graphic relationship between the length and width of rectangles. "Graphing the Area of a Rectangle" (page 83) gives students the opportunity to calculate and then graph the set of ordered pairs (l, w) for a rectangle with an area of 24 square units. By using a graphic representation, as well as traditional computation, this activity encourages students with a variety of learning modalities.

"The Valley on Mars" (page 87) is a delicious way to use nonstandard units of measure (such as Oreo cookies) to find the length of a valley on Mars that is thirteen times longer than the Grand Canyon. Just how expensive would it be to take this measurement using this tasty unit of measure?

Enjoy these motivating and enriching activities and projects with your students. Each allows students to touch and experience the mathematics of geometry and measurement.

Pennies and the Sears Tower

MATH TOPICS

measurement, computation, estimation, conversions, data collection and analysis, problem solving

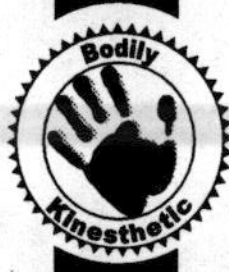

PRIOR KNOWLEDGE NEEDED

1. converting within English system of measurement
2. finding the arithmetic mean
3. ruler measurement

DIFFERENTIATION STRATEGIES

Principles

Flexible grouping: Students will work in groups of four to conduct this activity.

Ongoing assessment: Student activity sheet asks students to explain how they made their estimate, explain and show their calculations, and describe how the experiment was designed. This can be used as part of the assessment process.

Teacher's Strategies

Product: Tiered Journal Questions: Larger numbers are more difficult to work with than smaller ones. Journal question Level 1 asks students to estimate the number of pennies to reach their own height (approximately 5 feet). Journal question Level 2 increases the length of the distance to 2 miles.

According to Students

Learning Styles/Multiple Intelligences: logical/mathematical, bodily/kinesthetic, visual/spatial, verbal/linguistic, interpersonal

MATERIALS NEEDED

1. a bag of 20–25 pennies for each group
2. rulers

3. calculators
4. an overhead transparency of class data collection table

TEACHING SUGGESTIONS

Engaging the Students

Skyscrapers are an interesting topic, and most students find them fascinating. It is interesting that less than ten years ago, eight out of the ten tallest buildings were in the United States. In 2005, eight out of the ten were located in the Middle East and Asia. An interesting fact—the Empire State Building, which was completed in 1931, remains on the top 10 list of tallest buildings. It was the tallest building for the longest period of time (from 1931 to 1972) and has remained one of the tallest for over 70 years! Explain to students that today's experiment will be to "measure" one of the top ten tallest buildings in the world, the Sears Tower, in a rather unusual way—using a stack of pennies! Ask students, "How many pennies do you think would need to be stacked one on top of the other to be as tall as the Sears Tower?"

The Exploration

After estimates have been made and recorded, assign students to groups of four, and give each group a copy of the student activity sheet. You might want to have an overhead transparency of the activity sheet. Ask students to estimate how much they think a stack of pennies the height of the Sears Tower might be worth. Write all of these estimates down. If there is a great range of values, ask students why they think this occurred. Now read the problem with the students and brainstorm how an experiment might be set up that would help the class make a more reasonable estimate of the value of a stack of pennies that is 1450 feet tall.

Give each group a bag of pennies (containing 20 to 25 pennies) and let them work together to design an experiment to help them compute an answer to the problem. After the groups have completed the activity and answered the questions in the "Our Work" box, have the groups enter their solutions on the class data collection table.

Debriefing

As a group, analyze the data that each group has collected. Ask students: 1) Should all of the data be used or are there outliers that should be removed before an average or a mean is calculated? 2) If there is a variance between the data, why do you think that occurred? 3) What did your original stack of pennies look like, and why did you set up your experiment the way you did? 4) How did you convert the value of your stack from pennies to a more reasonable unit of measure (dollars)?

ASSESSMENT

1. Student products: The student activity sheet asks students to explain how they set up their experiment and to show and explain their calculations. This written part of the experiment can be used to help assess the group's level of understanding.
2. Journal questions:
   a. Level 1 question: Measure your height to the nearest $\frac{1}{2}$ inch. Write your height in your journal. Now use a stack of pennies to help you estimate how many pennies tall you are. Be sure to explain 1) how you set up your experiment, 2) how many pennies you used to make your estimate, and 3) how much your stack of pennies is worth in dollars.
   b. Level 2 question: Coin Middle School is 2 miles from its neighboring school, Currency Elementary. If you were to place pennies one on top of the other for a height of 2 miles, how much would your stack of pennies be worth? Explain how you set up your experiment.

VARIATIONS FOR A TIERED ACTIVITY

The Internet has many sites with information related to skyscrapers. Interested students can find diagrams of skyscrapers on-line and then draw each of the buildings to scale.

Pennies and the Sears Tower

Directions: The Sears Tower in Chicago is 1450 feet tall. If you were to stack pennies (one on top of each other) to the height of the Sears Tower, how much do you think your stack of pennies would be worth? ________________

To be more confident of the accuracy of our estimate, we can conduct an experiment.

Work in your group to

1) find how many pennies there would be in a stack 1450 feet high, and

2) find the value of this stack (using reasonable units of measure). Write your answers and show your work.

1. The number of pennies in a stack 1450 feet tall: ________________

2. The value of these pennies (using reasonable units of measure):

____

OUR WORK

Be sure to 1) explain how many pennies you used to make your estimate
2) describe and show your calculations
3) explain how you decided how you wanted to design your experiment

Pennies and the Sears Tower

CLASS DATA TABLE

Group	Number of Pennies	Value of Pennies ($)
Mean		

The Bouncing Ball

MATH TOPICS

data collection tables, measurement, averages, graphing

PRIOR KNOWLEDGE NEEDED

1. measuring accurately using a meterstick
2. graphing on the coordinate plane
3. labeling the x and y axes in appropriate and equal units
4. knowledge of the arithmetic mean

DIFFERENTIATION STRATEGIES

Principles

Flexible grouping: Students work in groups of four. These can be set up as homogeneous or heterogeneous groups depending on student needs.

Ongoing assessment: Observe students and use productive questions; data collection tables and graphs can be used for assessment.

Teacher's Strategies

Product: Journal question: Using a journal question helps meet the learning style needs of more students.

According to Students

Learning Styles/Multiple Intelligences: logical/mathematical, bodily/kinesthetic, interpersonal, verbal/linguistic, intrapersonal

MATERIALS NEEDED

1. one meterstick per group of four students
2. masking tape
3. calculators
4. rubber balls (one per group)
5. an overhead transparency of data collection table and graph

TEACHING SUGGESTIONS

Engaging the Students

Ask students, "Do you think that a ball dropped from a height of 100 cm will bounce back higher if it is dropped from a height of 40 cm?" Have students discuss what they think will happen in their groups and share their thoughts with the entire class. Explain that in this experiment, a ball will be dropped from heights of 100 cm, 80 cm, 60 cm, and 40 cm. The bounce-back height will be measured in three different trials and then the mean (or average) distance will be calculated for each height. Students will work in groups of four.

The Exploration

Each member of the group will be assigned a task. The tasks will rotate for each of the four heights so everyone has a chance to experience each job. The tasks are as follows:

1. Recorder: will be responsible for recording the bounce-back height for the three trials.
2. The Measurer and Dropper: will be responsible for measuring the correct height (either 40, 60, 80, or 100 cm) from which the ball is to be dropped.
3. The Observer: will be responsible for carefully noting the bounce-back height. If the three bounce-back heights differ greatly, it may be necessary to have a second member of the group help the Observer corroborate the height.
4. The Calculator: will find the arithmetic mean of the three trials.

Each group will need a meterstick and masking tape. The metersticks must be attached to a table at a 90° angle to the floor. Each of the trial heights should be recorded on the graph and a line drawn that represents the trend of the data.

Once the data is graphed, students can be asked to write a verbal description of the experiment along with their analyses of their results.

Debriefing

The debriefing can be done as a class discussion with each group being given the opportunity to describe their results and give reasons for their conclusions.

ASSESSMENT

1. Student products: Assess the accuracy of measurement and calculations on the data collection table and the accuracy of the graph. Were the units on the graph correct and was there a trend line to approximate a line of best fit?
2. Journal question: In every experiment, there are dependent and independent variables. Research both of these conditions, and describe each of the conditions of this experiment using these terms.
3. Productive questioning: As you observe each of the groups, the following questions can be asked and used as part of the assessment process: "Who in the group can explain why you made the predictions you did? Do your predictions appear to be correct? Have you had any difficulty determining the bounce-back height? If you did, how did you correct the problem? What effect did you note the drop height had on the bounce-back height? and so forth. The type of questions asked should be based upon the teacher's observations and any problems the students might be having.

VARIATIONS FOR A TIERED ACTIVITY

The experiment can be changed so that the ball types become the experimental variable. The height remains constant, and the type of ball varies. There should be three trials for each and an average taken. Students can be required to designate the dependent and independent variables in this revised experiment.

The Bouncing Ball

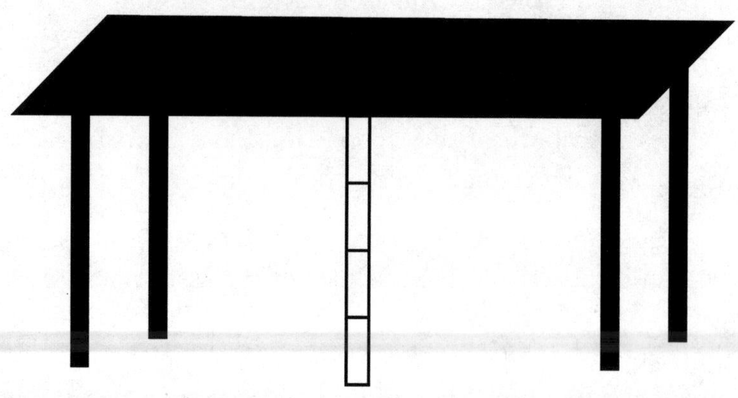

Directions: Tape a meterstick to the side of a table as shown above. Be sure that the meterstick is perpendicular to the floor. Drop the ball from each of the heights indicated on the table below three times and find the average height the ball bounces back on the first bounce. Read the height, as carefully as you can, from the bottom of the ball. After your group has completed the experiment, graph your results.

DATA COLLECTION TABLE

Height Ball Dropped 100 cm Trials	Height Ball Bounced (cm)	Average of Three Trials
Trial 1		
Trial 2		
Trial 3		
Height Ball Dropped 80 cm Trials	Height Ball Bounced (cm)	Average of Three Trials
Trial 1		
Trial 2		
Trial 3		
Height Ball Dropped 60 cm Trials	Height Ball Bounced (cm)	Average of Three Trials
Trial 1		
Trial 2		
Trial 3		
Height Ball Dropped 40 cm Trials	Height Ball Bounced (cm)	Average of Three Trials
Trial 1		
Trial 2		
Trial 3		

The Bouncing Ball

Directions: Label each axis using equal units. Then locate and graph your data. Do not connect the points, but draw a "line of best fit" or a line that best describes the trend of the data. Describe the trend of the data on the lines that follow.

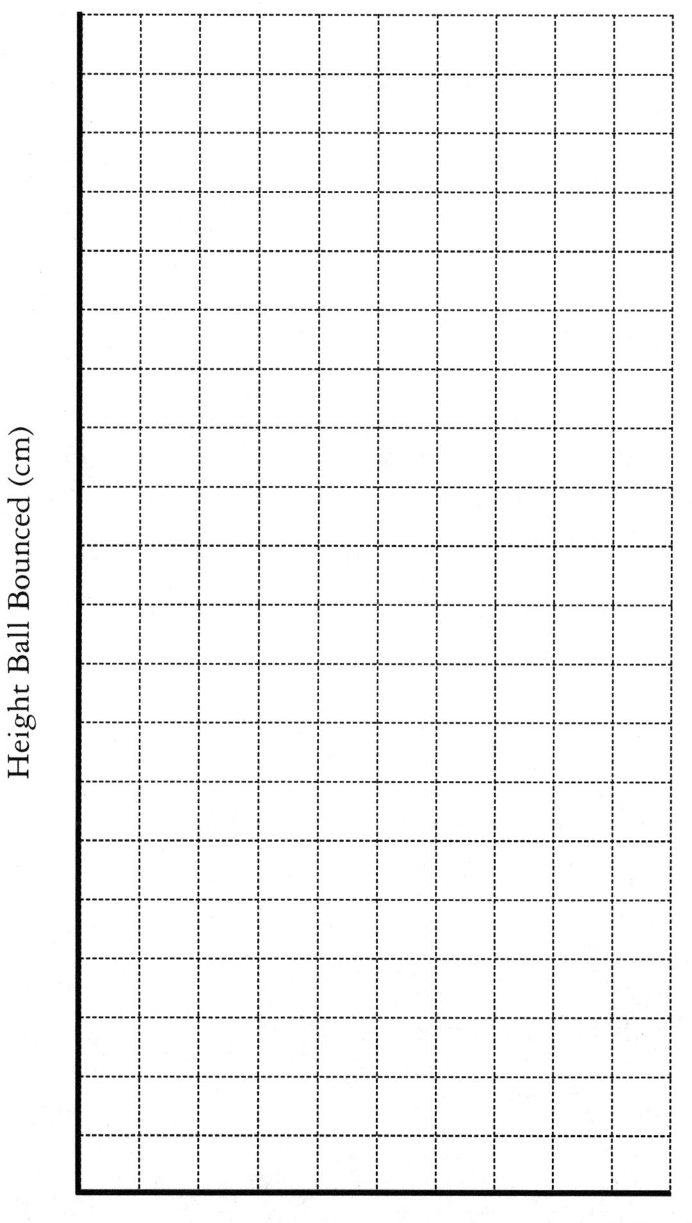

____

____

____

Soda Pop Math

Math Topics

geometry, spatial reasoning, problem solving, measurement

PRIOR KNOWLEDGE NEEDED

1. understanding how a two-dimensional net can be used to construct a three-dimensional polyhedron
2. making a scale drawing
3. taking accurate measurements using an English unit ruler

DIFFERENTIATION STRATEGIES

Principles

Flexible grouping: Students work in groups of four to collaborate on this construction.

Ongoing assessment: Throughout the construction process, measurements, design elements, progress, and participation can be assessed.

Teacher's Strategies

Product: Tiered Journal Questions: Journal question Level 1 asks the students to explain in their own words how knowing what the net of a cylinder looks like helps with other geometric measurements. Journal question Level 2 adds another dimension to the question; this one asks the students to find the surface area of a cylinder given certain measurements.

According to Students

Learning Styles/Multiple Intelligences: logical/mathematical, visual/spatial, verbal/linguistic, interpersonal, bodily/kinesthetic

MATERIALS NEEDED

1. half a sheet of tagboard for each group of four students
2. scissors
3. protractors
4. rulers and metersticks
5. construction paper or poster board
6. colored markers, crayons, and colored pencils

7. glue or Scotch tape
8. calculators and computers (if available for word processing)

TEACHING SUGGESTIONS

Engaging the Students

Help students understand geometric nets by starting with a more familiar figure, such as a cube. Discuss with students how a model of a cube can be made from the net of six squares.

Give students 1-inch graph paper (in the appendix) and scissors and have students work in pairs to construct a cube using six connected squares. While there are many possible configurations, this is one of them:

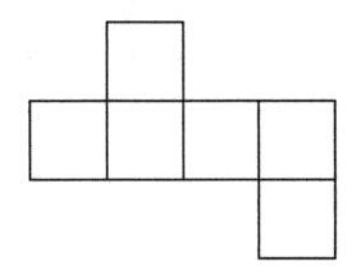

The Exploration

After students understand the concept of a net, discuss the geometric shapes that make up a cylinder. Ask, "What is the relationship between the sides of the cylinder (the rectangle) and the top and bottom of the cylinder (the circles)?" Work through the discussion that precedes the directions on the student activity sheet. A net of a cylinder would look like this:

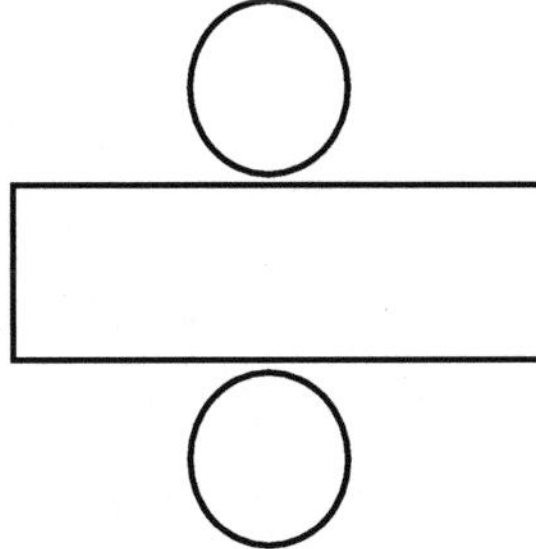

The length of the rectangle is equal to the circumference of the cylinder. The height of the rectangle is equal to the height of the cylinder.

Students can use the 1-inch graph paper to design this net for the cylinder. Precise measurement is very important—stress to students that they cannot have gaps or edges that overlap.

Debriefing

After all the groups have completed their "product cans," the cans can be displayed to the class. Group members can explain how their group completed the project and what each person's role was.

ASSESSMENT

1. Student products: Assess the analysis sheet that accompanies the activity using a rubric designed to assess all of the requirements of the cylinder

1) accuracy of construction, 2) quality of design, 3) completion of project, and 4) design of all of the sides as required by the project.

2. Journal questions:

   a. Level 1 question: Explain how understanding what geometric shapes a cylinder is comprised of would help you find the surface area and volume of this three-dimensional shape.

   b. Level 2 question: 1) Explain how understanding what geometric shapes a cylinder is comprised of would help you find the surface area and volume of this three-dimensional shape. 2) You have a product can that has the following dimensions: The top has a radius of $1\frac{1}{2}$ inches and a height of 3 inches. Find the surface area (in square inches) and the volume (in cubic inches) of this can.

3. TAP Activities: Students will choose from one of the projects listed below:

Task	Audience	Product
advertising professional	buying public	Create an advertisement to help sell the product that is in the cylinder you designed. Your advertisement must contain a picture of your group's product can drawn to scale.
songwriter	buying public	Write a song or a jingle to help sell the product that is in the cylinder you designed.
designer	mathematics students	Draw a collection of nets of other polyhedra (a minimum of two are required). Some examples are cube, pyramid, octahedron, and so forth. Draw them so that students will cut on the solid lines and fold on the dotted lines.
author	children 3 to 5 years old	Create a picture book about three-dimensional shapes. Each page should include an accurately drawn shape and the correct name of that shape. For example, a "can" should be labeled "cylinder."
an original idea presented by the student to be approved by the teacher		

Soda Pop Math

Directions: A soda pop can (cylinder) is made up of two basic geometric shapes. Can you name them?

If you took off the top and bottom of a can, cut it down the sides, and opened it up, what polygon would you have?

What would you call the length of this polygon (in relationship to the pop can)? the width of this polygon (also in relationship to the pop can)?

Work with your group to design a cylinder (can).

1. You have a sheet of tagboard that is 36 cm × 56 cm, or 2016 sq. cm.
2. The can needs to be constructed from one piece of paper. You cannot cut out individual pieces. This flat pattern is called a *net*. A net of a cylinder will be made up of a rectangle and two circles. Draw what you think a net of a cylinder will look like. Remember that you can only fold it to make it into the cylinder!

Our drawing of the net of a cylinder:

3. Your group is to design a net that becomes a cylinder when folded. Be sure to use as much of the tagboard as possible since any unused paper will be considered a waste of raw materials.
4. Decide with your group what product your can will hold, and make plans to design the outside accordingly.
5. What is on the outside of product cans? At the very least they have
   - the name of the product
   - some advertising to help sell the product
   - an attractive design to catch the buyer's eye
   - the amount of the product contained in the can (either by weight or volume)
   - nutritional information
   - perhaps a recipe to give the buyer more of a reason to purchase it
   - perhaps directions on how to prepare the contents
6. Be sure to put your designs and information on the outside of the can prior to its construction. It will be much easier to do your work on a flat surface than on a curved one!

Soda Pop Math

Let's analyze the size and shape of your product can (cylinder).

- How big was the sheet of tagboard you started with? ___________
- What are the dimensions of your design?

  a. height ___________

  b. diameter of top ___________

  c. radius of top ___________
- If you wanted to determine how much it would cost to design the surface of your can you would need to know the surface area. What is its surface area? ___________
- To inform the buying public of the amount of "stuff" your can will hold, you would need to know its volume. What is its volume? ___________

If it cost $0.003 per square cm to decorate the outside of the can (in order to attract the buying public), how much will you have to spend on this decoration? Explain how you got your answer.

In the space below, do all of your computations.

Paper-Folding Polygons

Math Topics

informal geometry, angle measurement, central angles, triangles, symmetry, problem solving

PRIOR KNOWLEDGE NEEDED

1. knowledge of central angles and interior angles of regular polygons
2. ability to follow oral directions

DIFFERENTIATION STRATEGIES

Principles

Flexible grouping: Students work individually on the activity, but directions are given to the whole class.

Ongoing assessment: Use attention-focusing, measuring, problem-solving, and reasoning questions during the paper-folding process. Finished products can be used for assessment.

Teacher's Strategies

Product: Journal question: Using a journal question as part of the assessment process gives students alternatives to traditional quizzes and tests.

According to Students

Learning Styles/Multiple Intelligences: bodily/kinesthetic, logical/mathematical, verbal/linguistic, visual/spatial, intrapersonal, naturalist

MATERIALS NEEDED

1. one sheet of paper ($8\frac{1}{2}$" × 11") for each student for each of the paper-folding activities
2. copies of the direction sheets for each student
3. scissors
4. colored markers, colored pencils, or crayons
5. construction paper

6. glue sticks or rubber cement
7. rulers

TEACHING SUGGESTIONS

Engaging the Students

These paper-folding activities can be done at the same time or spaced over a period of days or weeks. Since they are not dependent on one another, they can be done at any time. The questions asked during the paper-folding are basically the same for each activity. However, the size of the central angles will differ, depending on the shape, and the number of sections will vary.

Begin the activity by asking students, "What do we mean when we say a polygon is a regular polygon?" Ask students to draw a pentagon. Have students share their drawings. A comparison question might be, "What attributes do these shapes have in common, and how are the shapes different?" Draw the diagonals of one of the non-regular pentagons. Explain to students that the angles formed in the center of the pentagon are called the central angles, and the angles formed at the intersections of the sides are the interior angles. When the completed pentagon is unfolded, there will be 10 congruent central angles and 5 congruent sides. The number of degrees in each of the 10 central angles is 36°, but the number of degrees in each interior angle of a regular pentagon is 72°. When the hexagon is completed, there will be 12 central angles but only 6 congruent sides. The central angles of a regular hexagon are 60°; the interior angles are 120°. When the octagon is completed, there will be 16 central angles and 8 congruent sides. The central angles of a regular octagon are 45°; the interior angles are 135°.

The Exploration

Explain to students that they are going to start this project with a standard sheet of paper and end with regular polygons that can be made into unusual snowflakes. Explain that as the folding takes place, you will be asking them questions about the shapes and angles that are being formed. They must pay close attention and problem-solve the answers before they continue.

The more precise the folds, the more accurate the measurements will be. Step-by-step directions for folding are on the student pages. Students should be asked to measure or predict the number of degrees in the angles that are formed during the paper-folding process.

Ask students to design a shape on the folded pentagon (before they open it) and predict what it will look like. Examples of both are shown here:

When the folded pentagon is opened, the pentagonal design to the right is formed. It is an interesting activity for the students to analyze the shape before and after the pentagon is opened.

Debriefing

Give students a sheet of colored construction paper on which to glue their snowflakes. Ask students to describe the types of symmetry they see in their snowflakes. An attention-focusing question might be, "Look at the diagonals formed by the folds in your snowflake. Does there appear to be a relationship between these diagonals and the lines of symmetry?"

ASSESSMENT

1. Student products: Assess snowflake designs.
2. Journal question: Explain the difference between the interior angles and central angles of the pentagon. What must the sum of the central angles of any polygon be and why?

Paper-Folding a Pentagon

1. Fold a rectangular sheet of paper in half, as shown.

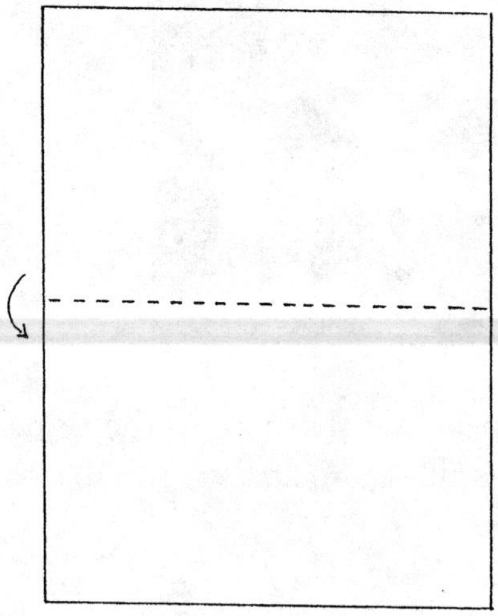

2. With the folded edge on top, fold in half again. Open it back up.

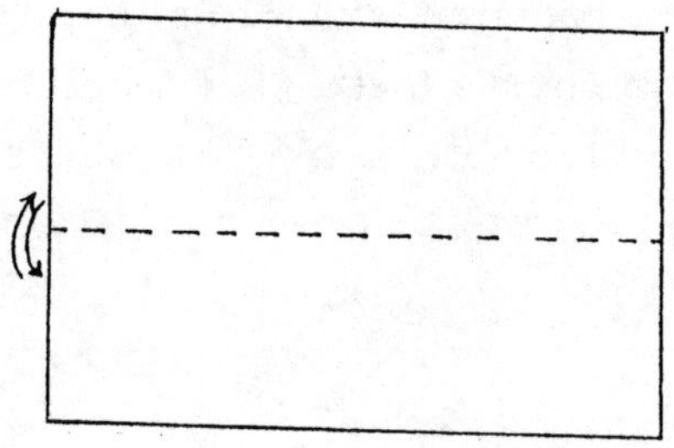

3. Label point P on $\overline{BC}$. Fold vertex A to point P.

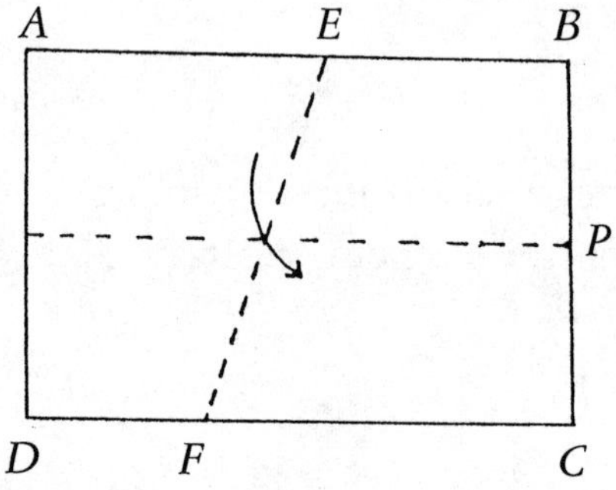

4. Fold $\overline{EF}$ to $\overline{EP}$. You will be bisecting $\angle E$.

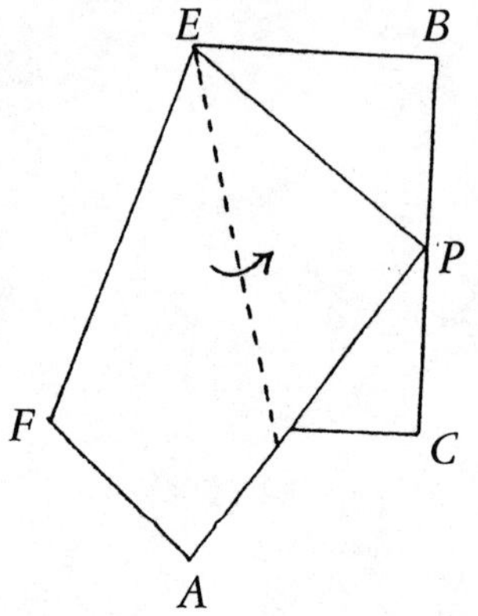

5. The shape now looks like this.

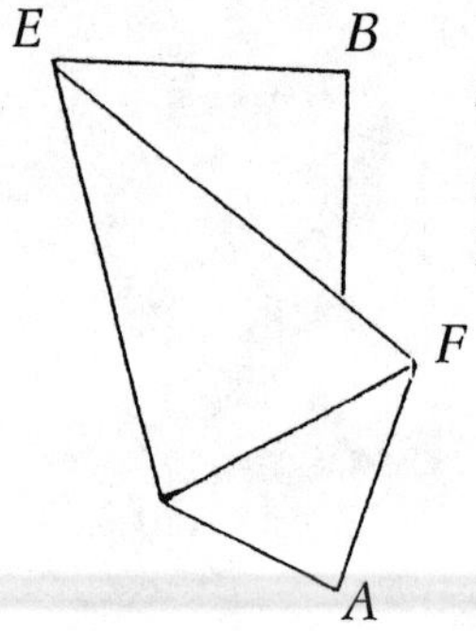

6. Turn over—front to back. Fold back vertex B as shown again bisecting $\angle E$.

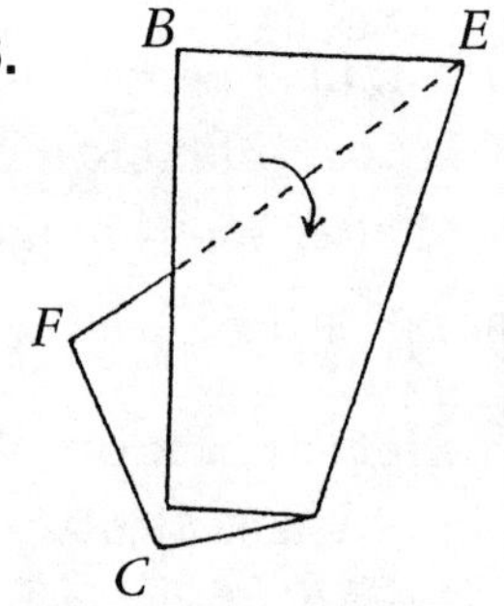

7. This is the final shape. $\angle E$ is 36°. Cut where shown.

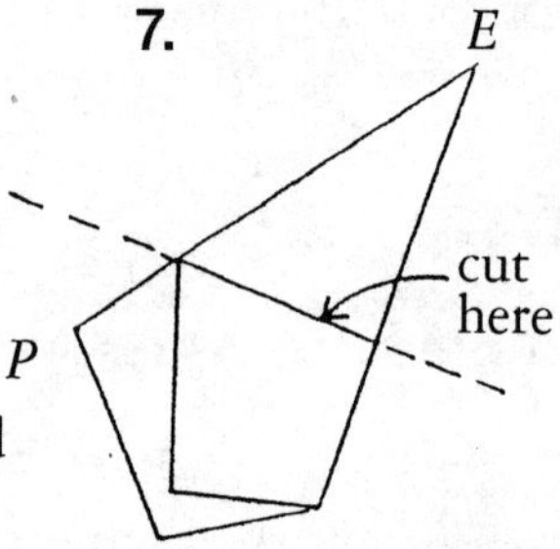

8. Open up to form a regular pentagon.

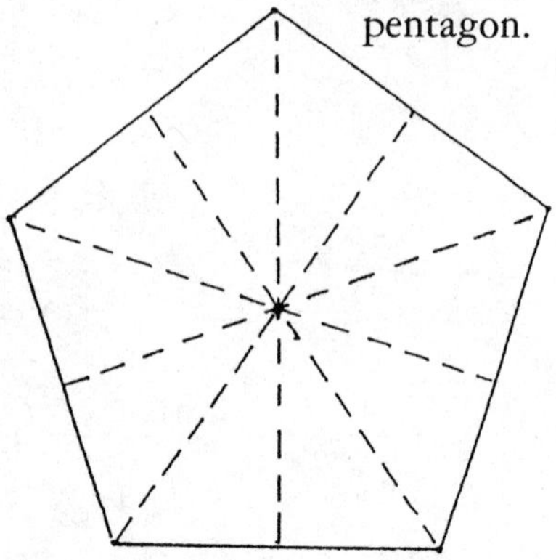

Paper-Folding a Hexagon

1.

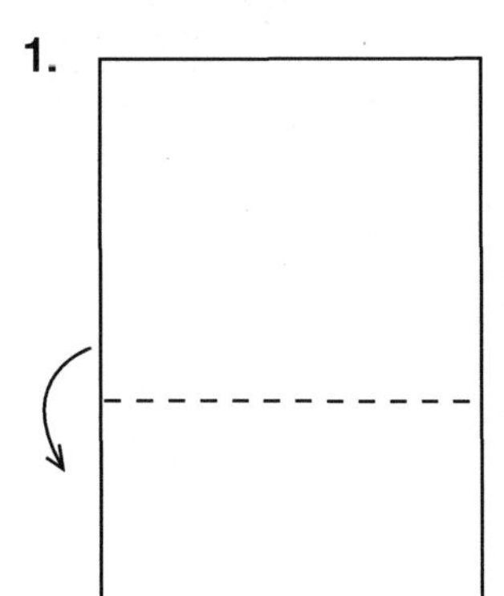

Fold a rectangular sheet of paper in half, as shown.

2.

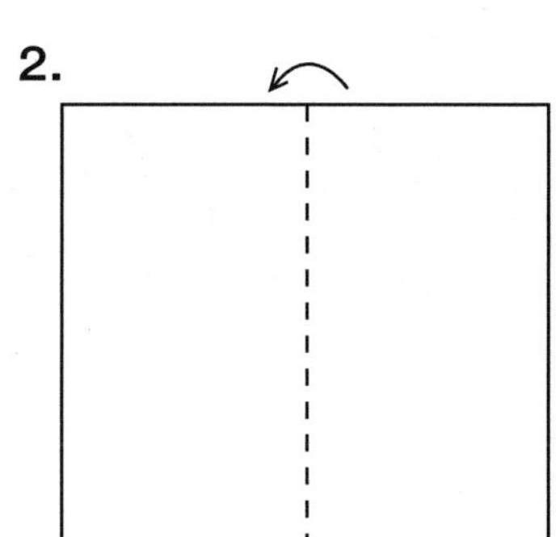

With the closed folded edge on top, fold in half again, as shown.

3.

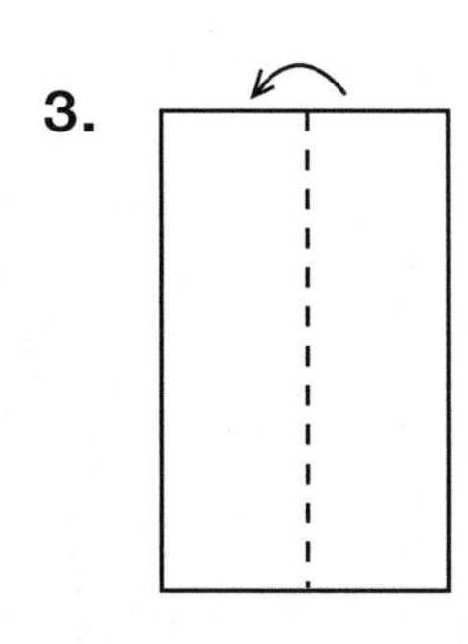

With the closed folded edge on the right, fold in half again, as shown, and open up!

4.

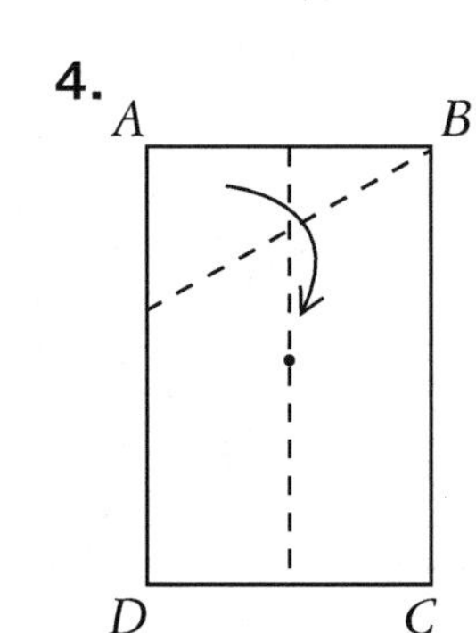

Fold vertex *A* so that it meets fold line.

5.

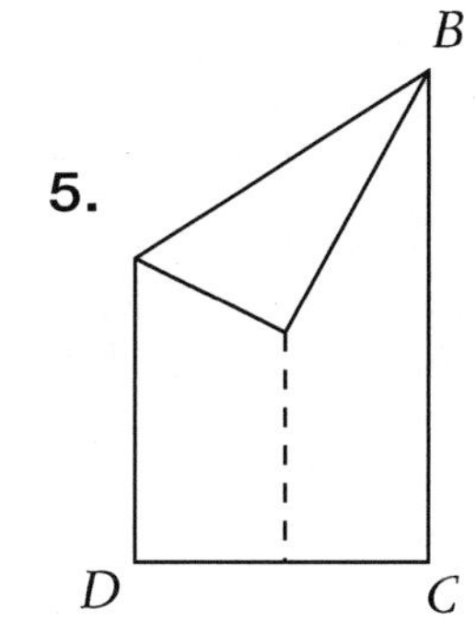

Fold a line from *B* to *D* behind; you are bisecting the angle at point *B*.

6.

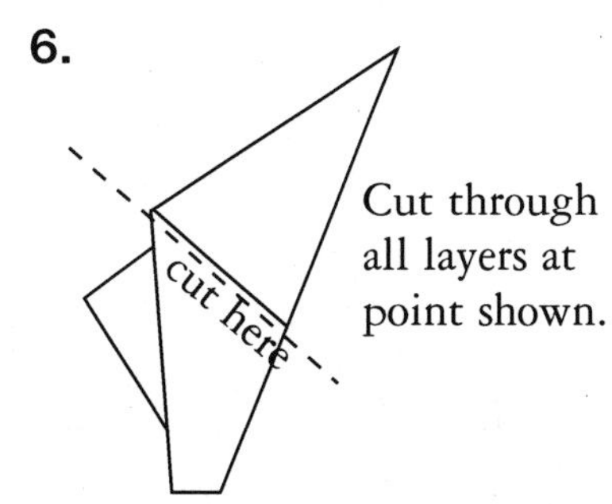

Cut through all layers at point shown.

7.

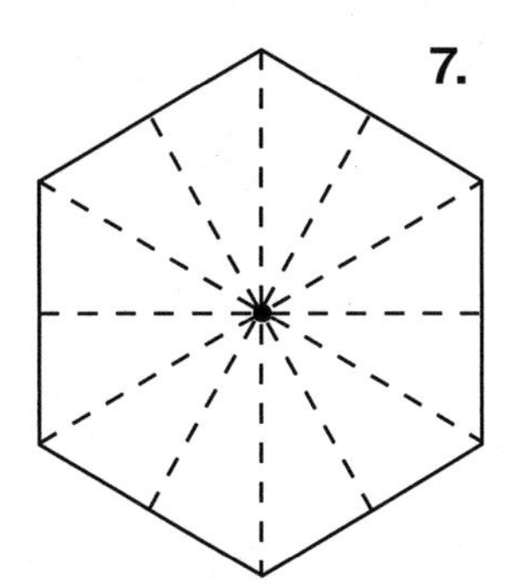

When unfolded, you have a hexagon. Refold, cut decoratively, to form a snowflake.

Paper-Folding an Octagon

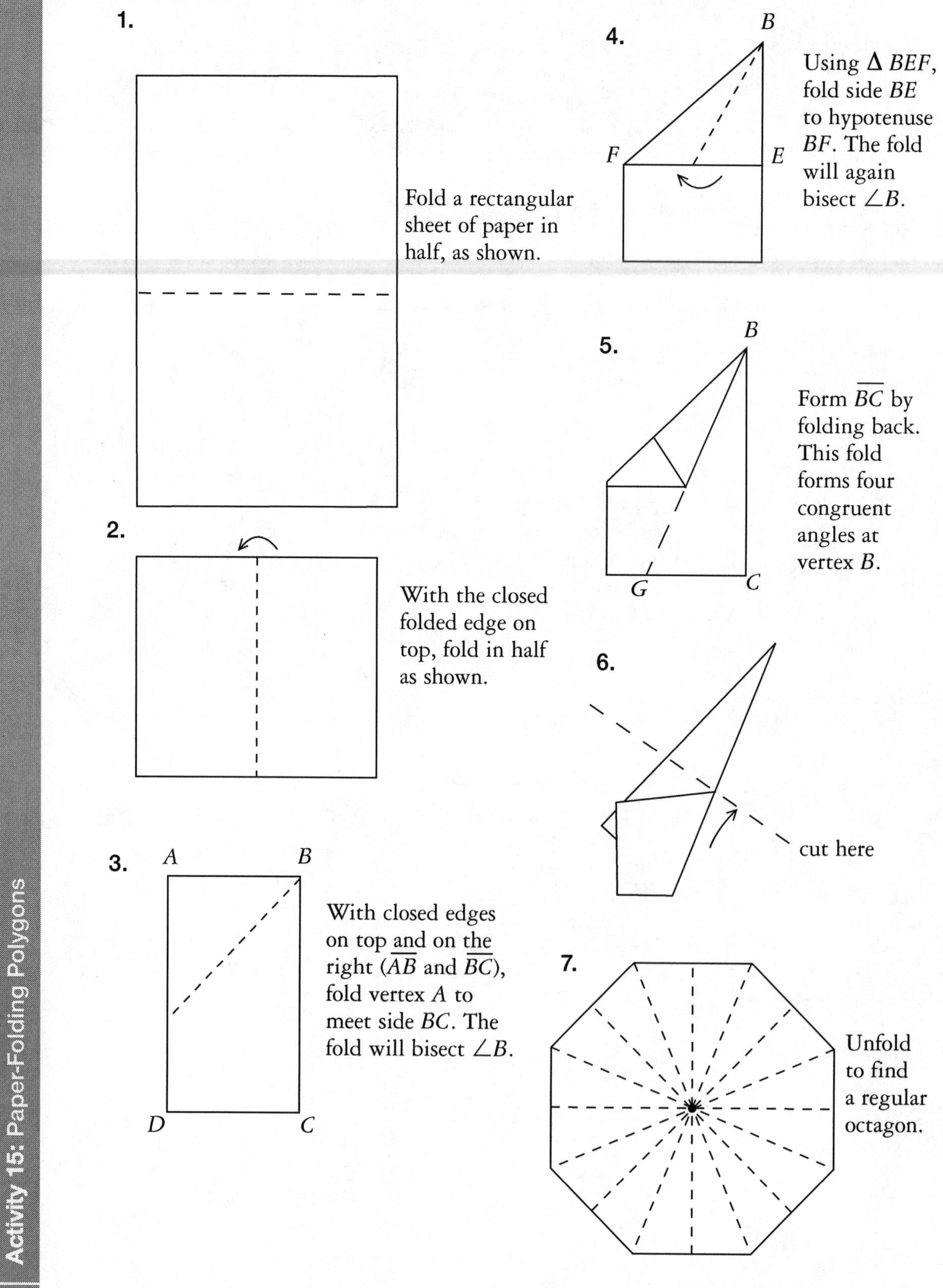

1.
Fold a rectangular sheet of paper in half, as shown.
2.
With the closed folded edge on top, fold in half as shown.
3.
A
B
D
C
With closed edges on top and on the right ($\overline{AB}$ and $\overline{BC}$), fold vertex A to meet side BC. The fold will bisect $\angle B$.
4.
B
F
E
Using ΔBEF, fold side BE to hypotenuse BF. The fold will again bisect $\angle B$.
5.
B
G
C
Form $\overline{BC}$ by folding back. This fold forms four congruent angles at vertex B.
6.
cut here
7.
Unfold to find a regular octagon.

Graphing the Area of a Rectangle

MATH TOPICS

area, graphing nonlinear equations, analysis of data using graphs, rational numbers

PRIOR KNOWLEDGE NEEDED

1. finding the area of a rectangle
2. graphing on the coordinate plane

DIFFERENTIATION STRATEGIES

Principles

Flexible grouping: This lesson is designed for whole-group instruction.

Ongoing assessment: Use productive questions while students work on the problem. Make sure students successfully complete the student activity sheet.

Teacher's Strategies

Product: Tiered Journal Questions: Both journal questions Level 1 and Level 2 ask students to describe in their own words a real-world situation that might have a graph such as the one formed by the two variables in the problem. However, journal question Level 2 asks students to define the dependent and independent variables.

According to Students

Learning Styles/Multiple Intelligences: logical/mathematical, visual/spatial, verbal/linguistic, intrapersonal, naturalist

MATERIALS NEEDED

1. one activity sheet for each student
2. calculators
3. an overhead transparency of the activity sheet to introduce the problem

TEACHING SUGGESTIONS

Engaging the Students

Introduce the activity to students using a rectangle with a constant area of 12. Ask students, "If the length of the rectangle is 1, what must the width be for the area to be 12?" Continue using all of the whole numbers 1 through 12. (The numbers on the table have been rounded to the nearest tenth.)

Length	1	2	3	4	5	6	7	8	9	10	11	12
Width	12	6	4	3	2.4	2	1.7	1.5	1.3	1.2	1.1	1

Have students discuss
a) what they predict the shape of the graph will be.
b) how the width is found when the length is known.
c) what the product of the two factors would be if the numbers had not been rounded.

The Exploration

Give each student a copy of the activity sheet. Have students problem-solve the various widths and then use the lengths and widths as x and y coordinates to form their graph.

Debriefing

On the activity sheet, students are asked to predict the shape of the graph, and then to describe the shape of the data. Ask students, "Did you accurately predict the shape of the graph or were there some surprises?" In all likelihood, students probably thought the graph would be a straight line. Give students time to problem-solve why their graphs looked the way they did. Ask them, "If a rectangle has a perimeter of 24, what are the possible side lengths? If the length and width of a constant perimeter were graphed, what do you think the graph would look like?"

ASSESSMENT

1. Student products: Review accuracy of calculations and graph of coordinate pairs.

2. Journal questions:

   a. Level 1 question: When you graphed coordinates formed by the length and width of a rectangle with a constant area of 24, the graph was a curved line. Give an example of a real-world situation in which ordered pairs would form the same type of graph.

b. Level 2 question: When you graphed coordinates formed by the length and width of a rectangle with a constant area of 24, the graph was a curved line. Give an example of a real-world situation in which ordered pairs would form the same type of graph. What are the dependent and independent variables of your problem?

VARIATIONS FOR TIERED ACTIVITIES

Allow some students to work in pairs to calculate and graph the area of equilateral triangles with sides of various lengths.

Graphing the Area of a Rectangle

Directions: The area of a rectangle is 24 square units. The width of the rectangle varies as the length changes. Complete the table below to show the inverse relationship between the length and width of this rectangle.

Length	1	2	3	4	5	6	7	8	9	10	11	12	13	14	15	16	17	18	19	20	21	22	23	24
Width	24	12		4.8																				1

Use the data to predict what shape the graph of this function will look like.

Now plot the points. Use this space to describe the actual shape of the data.

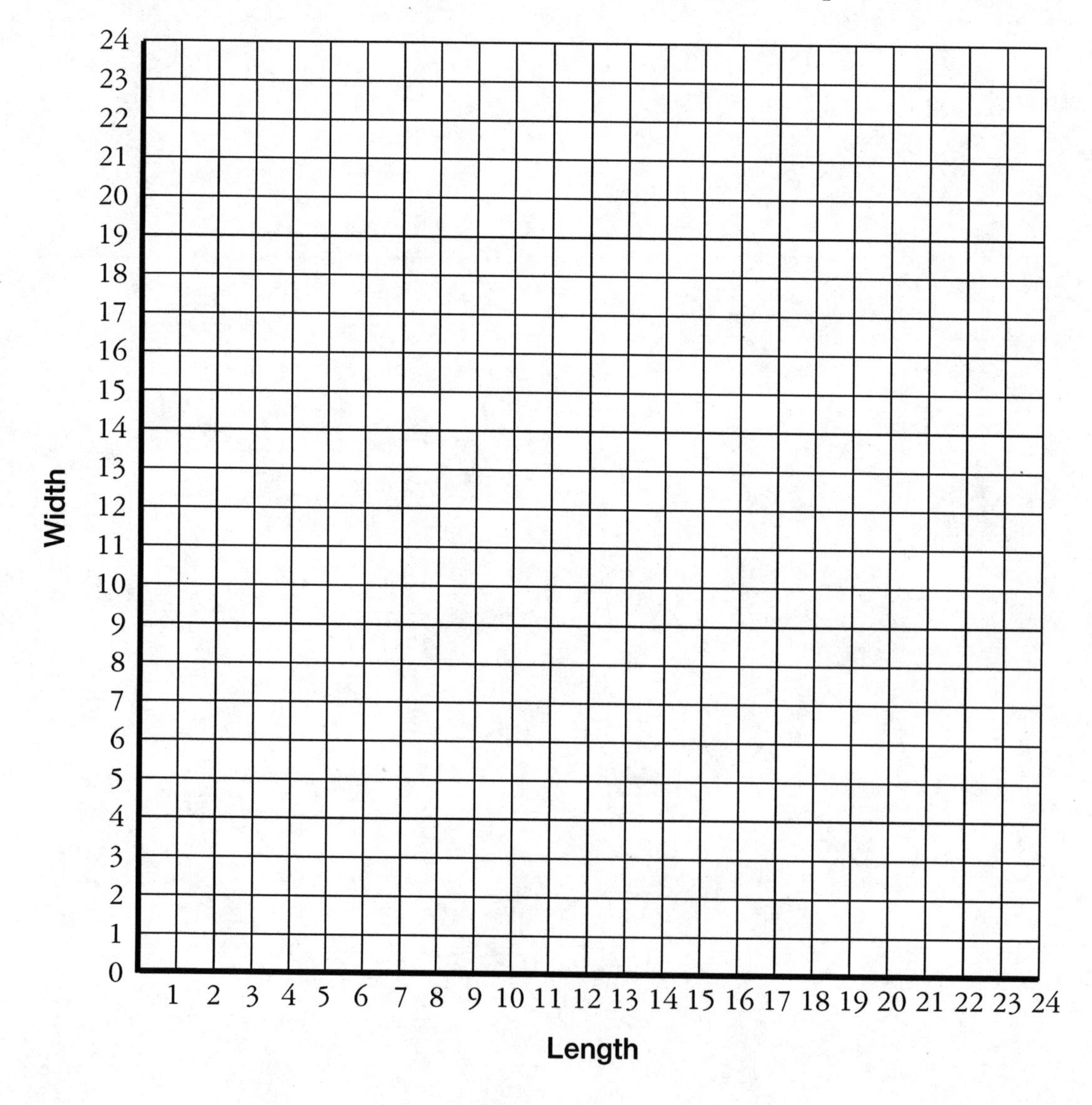

The Valley on Mars

MATH TOPICS

measurement, estimation, appropriate units of measure, computation, averages, conversions, open-ended problem solving

PRIOR KNOWLEDGE NEEDED

1. metric conversions (cookies/cm to cookies/km)
2. finding the arithmetic mean (average)

DIFFERENTIATION STRATEGIES

Principles

Flexible grouping: Students will work in groups of four to conduct this measurement experiment. The debriefing will be done with whole-group questioning and discussion.

Ongoing assessment: While groups are working, productive questioning can be used to keep them on track. The class data sheet can be used as part of the assessment process.

Teacher's Strategies

Product: Tiered Journal Questions: Journal question Level 1 asks students to measure the length and width of any classroom in the school and find the number of cookies it would take to cover the floor. For this question, students can use data from the experiment. Journal question Level 2 asks students to analyze the data collected using statistics.

According to Students

Learning Styles/Multiple Intelligences: logical/mathematical, bodily/kinesthetic, interpersonal, verbal/linguistic, visual/spatial

MATERIALS NEEDED

1. five cookies for each group of four students (Oreo cookies work well because they are quite uniform in diameter.)
2. metric rulers
3. calculators
4. one student activity sheet for each group
5. an overhead transparency of class data collection sheet

TEACHING SUGGESTIONS

Engaging the Students

Discuss with students the concept of unit of measure. Ask students to name some standard units we use to name our measurements. They might suggest inches, feet, meters, kilometers, and so forth. Ask if they have ever used any less traditional or non-standard units, such as the size of their foot. Very often young children will measure length using paper clips, or the weight of an object by comparing it to another object (it is heavier than this apple). Most likely, students have been using standard measurements for some time.

The Exploration

Distribute the class data collection sheets (one to each group) and ask students to predict how many "cookies" long the Mariner Valley on Mars might be. These estimates should be written on the data sheets so they can be referred to at the end of the lesson.

It is a good idea for students to measure more than one cookie, because all cookies are not exactly the same size. Once they have their measurements, they will need to convert in the following manner:

cookies/cm → cookies/m → cookies/km → cookies/4537 km

To calculate the cost of the cookies, students will need to know the cost of one package of cookies. Surprisingly, it will cost hundreds of millions of dollars (depending on the cost of the bag or box you supply).

Debriefing

Once student groups have collected their data and entered it on the class data collection sheet, discuss the results. Students can look for consistency of results, outliers (data that is so far from the mean that using it to find the mean or average would skew the result), and so forth. Students working on journal question Level 2 will need to have a copy of the class data collection sheet to answer the question.

ASSESSMENT

1. Student products: Group data can be used as part of the assessment process.
2. Journal questions:
   a. Level 1 question: Measure the length and width of any classroom in the school and, using the data you collected in "The Valley on Mars" activity, find the number of cookies it would take to cover the floor.
   b. Level 2 question: Use the class mean for "The Valley on Mars" activity to calculate range, median, mode(s), variation from the mean, and other statistical information you think might be interesting. Explain whether you think the data collected was of any value. Why or why not?

The Valley on Mars

The Grand Canyon in Colorado and Arizona is about 349 kilometers long, but Mars has the longest canyon in the solar system. It is called the Mariner Valley and is 13 times longer than the Grand Canyon. It is 4537 kilometers long, and it would stretch from the east to the west coast of our country.

What if we placed round cookies end-to-end and used these cookies as our unit of length? How many "cookies" long is the Mariner Valley on the planet Mars? How much would all of these cookies cost?

Work with your group to solve this problem.

1. Use your five cookies to help you solve this problem.
2. Remember, you are measuring in centimeters but must convert these measurements to kilometers to find the answer to this problem.
3. Calculate the cost of all of these cookies.

Our work

The Valley on Mars

CLASS DATA

Group	Number of Cookies	Cost of Cookies
Mean		

Were the class estimates accurate? If not, why do you think it was so difficult to estimate the number of cookies or their cost? Discuss these questions with your group and assign a spokesperson to explain your thinking.

Brush Up Those Skills Chapter 3

Directions: Roll one die; your group will work on the activity that matches the number on the die. If you have completed that activity, roll the die until it lands on an activity you have not done. On the next page, check off the activity your group completed. Make sure each group member has signed his or her name.

1	Take two sheets of paper, each $8\frac{1}{2} \times 11$ inches. Roll one to form a short cylinder (without a base). Tape it so that the edges just meet. Take the other sheet of paper and rotate it to form a tall cylinder. Tape it carefully. Observe the two cylinders to help you answer these two questions: 1) Do you think the two cylinders have the same volume? 2) If you do not think they do, which one do you think has the greater volume, the short cylinder or the tall cylinder, and why? With your estimates, calculate the volume of each cylinder. Use these formulas: $C = 2\pi r$ and $V = \pi r^2 h$. Record the volume of each cylinder and discuss the relationship between the two.
2	A B C D E F G H I J K L M N O P Q R S T U V W X Y Z Which of these letters have a vertical line of symmetry? ________________ have a horizontal line of symmetry? ________________ are symmetric around a point? ________________
3	Carefully draw a rectangle (use a protractor and a ruler) that is 3 inches × 5 inches. Use your ruler to draw a diagonal. Now cut it on the diagonal. 1. Find all the polygons that can be formed by placing congruent sides together. 2. Draw each polygon in your journal. 3. Describe each of the polygons with as much detail as you can.
4	On a Metric Scavenger Hunt, your group is to find something about 4.5 cm long, something that weighs between 1 and 2 kg, a container that holds about 100 ml, something with about a 25 cm^2 area. You may not use measurement devices for the hunt. After the hunt, measure each item. On a table, list the objects you had to find, what your group found, the actual measurements of each object, and the percent of error. Find the average or mean percent of error.
5	Use a protractor and a compass to draw a clock face. Label the center of the circle. There are 12 numbers evenly spaced around the clock. The number of degrees between each hour is ___°. If there are four people in your group, each person should draw and measure the angles formed by 3 different hours of the day. Use a protractor to measure the angle of each hour. Label each angle.
6	This is a picture of 4 congruent circles and 1 square. Each circle has a radius of 3 inches. Find the area of the unshaded portion. Explain how you solved this problem. Show all your work.

Brush Up Those Skills
Chapter 3

Our Progress Completing These Activities

We have completed | **Signatures of Group Members**

☐ Activity 1 ______________________
____
____
____

☐ Activity 2 ______________________
____
____
____

☐ Activity 3 ______________________
____
____
____

☐ Activity 4 ______________________
____
____
____

☐ Activity 5 ______________________
____
____
____

☐ Activity 6 ______________________
____
____
____

Data Analysis, Statistics, and Probability

Statistical thinking will one day be as necessary for efficient citizenship as the ability to read and write.

—H. G. WELLS

For students to understand the importance of statistics and data collection in our information age, activities and projects should be designed that model how statistics are used in the real world. All too often, students are asked to analyze textbook data and solve meaningless problems. To develop worthwhile experiences for students, teachers must investigate the reasons data is collected. In the real world, we collect data to make predictions, help explain patterns or trends, answer questions regarding preferences, or help explain natural phenomena. The data collection projects contained in this chapter help students relate to these real-world applications.

The probability activities and projects in this chapter focus on hands-on, real-world applications of probability. The first relates to geometric area, and the second focuses on sample space. But each makes connections between school mathematics and math in the real world.

The first activity, "Predicting Colors in a Bag of M&M's" (page 95), involves students in a very tasty experiment that helps demonstrate the usefulness of data collection and analysis in making predictions. How confident would anyone be in guessing how many of each color M&M's there are in a "mystery bag"? But after collecting some data, students have additional facts to help them make an informed estimate rather than an uninformed guess!

"Vowels, Vowels, Everywhere" (page 102) combines mathematical poetry and data collection and analysis to help students better understand the frequency of vowels in the English language. Students also learn how data helps explain patterns and trends.

"Phone Home" (page 108) is an amusing data collection activity that uses poetry to help students experience the speed of light, a natural phenomenon, and explain the speed in concrete terms. Miranda the Martian is homesick and sends a message home. How long will it take, at the speed of light, to reach her loved ones? Using stopwatches and careful data collection, students see how data can be used to explain natural phenomena.

When you fill out a form that contains little squares in which to put information, have you ever wondered how the designers of the form knew what would be an appropriate number of boxes? How do they know how many letters there are in your first name? Why do most of us not run out of boxes? The activity "How Long Is Your First Name?"

(page 112) uses statistics and deviation from the mean to help us answer this question. It is a motivating approach to explain patterns or trends.

The activity "Geometric Probability: Dartboards and Spinners" (page 117) asks students to consider the fairness of a dartboard. Are all games of skill and chance fair? How do we know if a game is fair? Using geometric probability, students calculate theoretical probability and design a spinner to conduct their own experiments.

Students have a hands-on opportunity to experiment with probability and sample space when they play "Flipping Three Coins: Heads or Tails" (page 125). After first hypothesizing their results, students toss three coins and then record their results in a frequency table. After conducting the experiment, they reevaluate their original hypothesis. By finding the sample space, students are able to compute the theoretical probability. This is a hands-on approach to a very abstract concept!

Predicting Colors in a Bag of M&M's

MATH TOPICS

data collection, organization, and analysis; box-and-whisker plots; percent of difference; absolute value; problem solving; critical thinking

PRIOR KNOWLEDGE NEEDED

1. calculating percent of difference
2. finding the median of data

DIFFERENTIATION STRATEGIES

Principles

Flexible grouping: Students work on this activity as individuals. During the debriefing session, students will work in pairs, and there will be whole-group instruction.

Ongoing assessment: Box-and-whisker plots designed by students can be used as part of the assessment process. While students are designing the box-and-whisker plots, teachers can observe students and ask attention-focusing, problem-posing, or reasoning questions.

Teacher's Strategies

Product: Tiered Journal Questions: Journal question Level 1 is easier because it contains fewer pieces of data. Journal question Level 2 has more data and asks students to analyze a box-and-whisker plot.

According to Students

Learning Styles/Multiple Intelligences: logical/mathematical, verbal/linguistic, interpersonal, bodily/kinesthetic

MATERIALS NEEDED

1. one bag of snack-size M&M's for each student, and one extra to serve as the "mystery bag"
2. an overhead transparency and student copies of each of the student activity sheets
3. calculators

TEACHING SUGGESTIONS

Engaging the Students

Hold up the "mystery bag" of M&M's, and ask students the question posed on the first activity sheet, "How confident would you be in predicting the number of each color of M&M's in this particular bag? We are not interested in the total number of candies in the bag, but the number of each color." As students guess numbers, ask them again, "Just how confident are you in your guesses?" Explain that in the real world, data are collected and analyzed to help us answer questions like these. By having additional information, we are able to make better estimates, more educated "guesses," and increase our levels of confidence.

The Exploration

Each student counts the number of each color of M&M's in his or her bag and supplies that data to the class data table. Once the data are collected, the steps below should be followed to construct the box-and-whisker plots. (A sample box-and-whisker plot is shown.)

1. Order the data for each color from least to greatest.
2. Find the median of the data; indicate this on the appropriate spot on the number line.
3. Find the median of the lower half of the data (this is called the lower quartile); indicate this on the appropriate spot on the number line.
4. Find the median of the upper half of the data (this is called the upper quartile); indicate this on the appropriate spot on the number line.
5. Use these three median scores to form the "box."
6. Place a dot on the least and greatest data, and connect them with a line to form the whisker.

For example, suppose the data for a class of 20 organized from least to greatest were as follows:

0, 1, 1, 2, 2, 2, 3, 3, 4, 4 5, 5, 7, 7, 7, 8, 8, 8, 9, 9

The median (or middle score) is between 4 and 5, so the median is 4.5. The lower quartile is 2, and the upper quartile is 7.5. These are indicated on the number line in the following way:

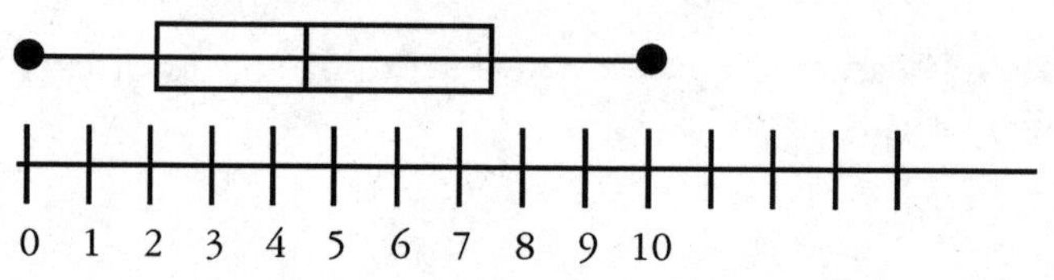

The range of the data is between 0 and 9; the median is 4.5. Fifty percent of the data falls between 2 and 7.5 (in the box). While students still have a range of about 5 to make their predictions, the data collection and graph help them narrow down their options. For example, the prediction can be 4 ± 2 (meaning that anything between 2 and 6 of the color would be considered correct). Looking at the data, this is a reasonable estimate.

Use the data in the class data table to draw a box-and-whisker plot with the students for this set of data. Go through the steps outlined above. Ask students the reasoning questions such as, "How do we know what the range of the data is? What percent of the data is in the "box"? What percent of the data is in the lower quartile? In the upper quartile?" If it appears that students are still confused about the design of the box-and-whisker plot, go through, step-by-step, making a plot for the color orange.

Debriefing

Students will need to use the class data to draw the box-and-whisker plots for the remainder of the colors. Ask students to predict using their plots (±1 or 2) how many M&M's of this color are in the mystery bag. In this instance, students might predict 4 (±2)—any number between 2 and 6 would be considered a correct guess. Write all of their suggestions down, and then have the class agree on a number—there should be a consensus. Write these predicted numbers in the correct column of the frequency table on the next page.

After predictions have been made for all of the colors, take out the mystery bag of M&M's. Write down the actual number of each color in this mystery bag and whether it was correct to ±2. Have students work in pairs to answer the analysis questions on the activity sheet.

ASSESSMENT

1. Student products: Individual student box-and-whisker plots and the data analysis sheet prepared by each pair of students can be used as part of the assessment process.
2. Journal questions:
    a. Level 1 question: This table shows the fuel economy of ten cars manufactured by Toyota in 2005. The mileage is in miles per gallon of city driving. Place the data in order from least to greatest, and use it to design a box-and-whisker plot. Be sure to explain how you solved this problem.

4Runner	RAV4	Tundra	Avalon	Sequoia	Camry	Echo	Highlander	Matrix	Prius
18	24	16	22	15	20	35	22	26	60

Source: www.fueleconomy.gov/feg/bymake/Toyota2005.shtml

b. Level 2 Question: This table shows the average life span of twenty-six animals. The data is listed from longest to shortest, but you will need to organize it to find the four quartiles and the range of the data. All data is in years.

Box turtle	100	Bactrian camel	12
Human	80	Domestic cat	12
Asian elephant	40	Domestic dog	12
Grizzly bear	25	Leopard	12
Horse	20	Giraffe	10
Gorilla	20	Pig	10
Polar bear	20	Squirrel	10
White rhino	20	Red fox	7
Black bear	18	Chipmunk	6
Lion	15	Rabbit	5
Lobster	15	Guinea pig	4
Rhesus monkey	15	Mouse	3
Black rhino	15	Opossum	1

Source: worldalmanacforkids.com/explore/animals4.html

Use your box-and-whisker plot to answer these questions:

1. Does there appear to be an outlier in this set of data? Is there data that is very far from the mean?
2. What is the range of the data?
3. What is the range of the middle 50%? of the lower quadrant? of the upper quadrant?
4. What do these ranges tell you about the data?

Predicting Colors in a Bag of M&M's

How confident would you be in predicting the number of each color of M&M in this particular bag? How many blue or red or orange M&M's do you think you would find? Do you think the same number of each color would be found in each bag?

By collecting some data, we can narrow our choices and more accurately predict or estimate what we might find in a "mystery bag" of M&M's.

Directions: Count how many of each color you have in your snack-size bag of M&M's. Be sure to write the numbers in the chart, as you will need to report them so they can be added to the class data.

CLASS DATA

Color	Student Data: Numbers of Each Color																									
Red																										
Orange																										
Green																										
Blue																										
Brown																										
Yellow																										

Let's use the data we have collected to make one box-and-whisker plot. This type of graph will help us make more accurate predictions. First, we must take the data and order it from least to greatest. Then we use this ordered data to form a box-and-whisker plot.

Red:

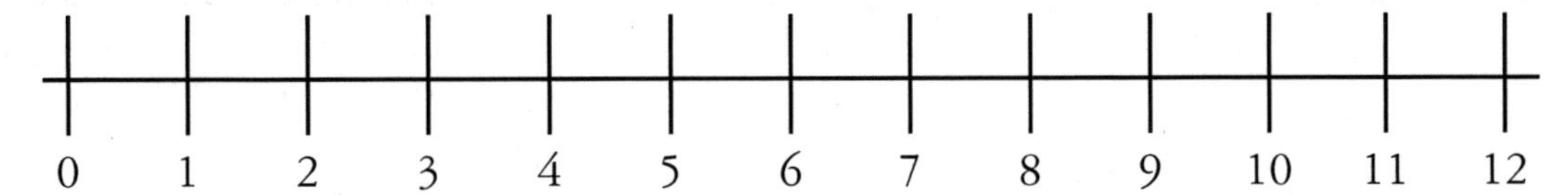

Predicting Colors in a Bag of M&M's

Orange:

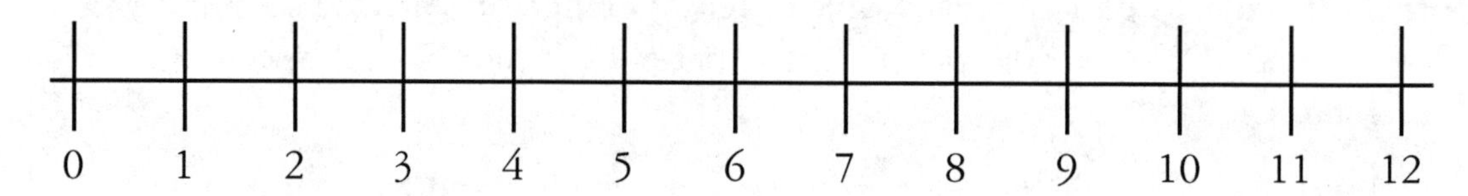

Green:

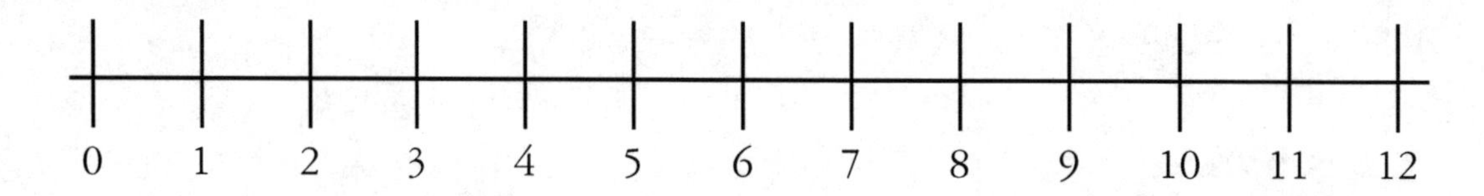

Blue:

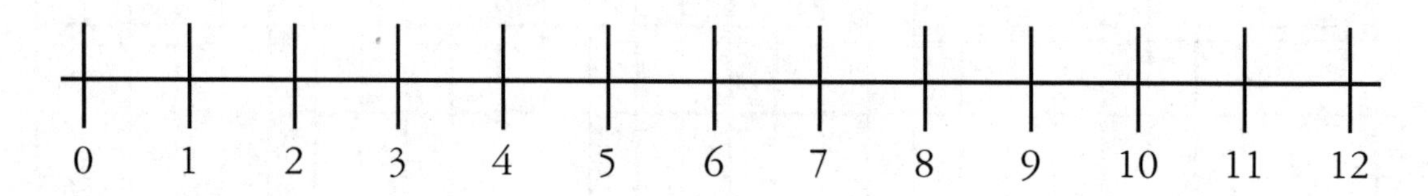

Brown:

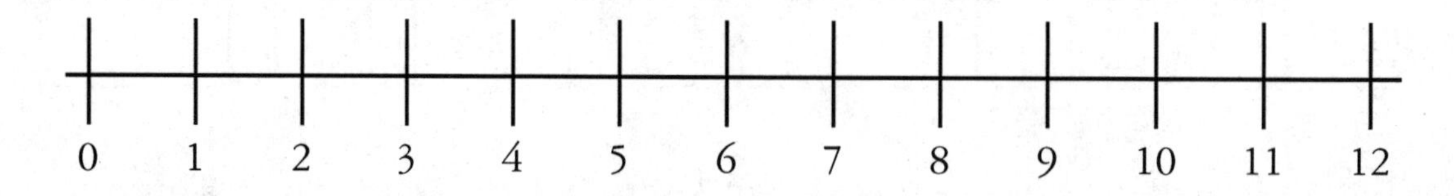

Yellow:

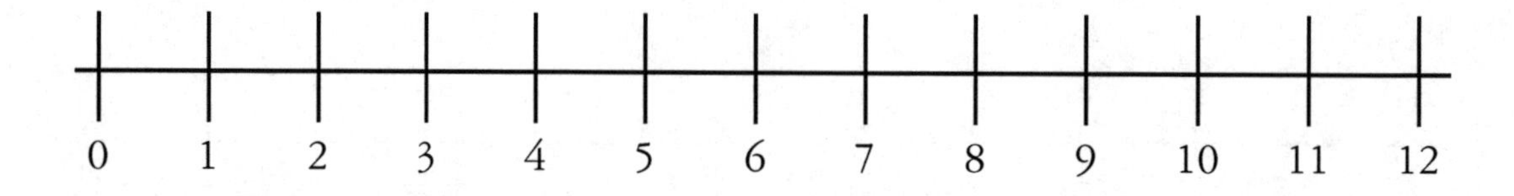

Predicting Colors in a Bag of M&M's

Now let's use the results of the data collection and our box-and-whisker plot graphs to make our predictions. How many of each color will there be in our "mystery bag"?

Color	Prediction	Actual Count	Correct/Incorrect (to ± 2)
Red			
Orange			
Green			
Blue			
Brown			
Yellow			
		% Correct	

1. Do you think the data collection and box-and-whisker plots helped make more accurate predictions? Why or why not? Explain your answer.

   ____

   ____

2. The Mars Candy Company says that they manufacture certain percentages of each of the colors. Do you think that the same percentage of each color is manufactured? Why or why not? Explain your answer.

   ____

   ____

   ____

3. Can you think of other predictions that we could make by collecting data in this manner and then using box-and-whisker plots to find medians and ranges in the data? Describe one of these experiments.

   ____

   ____

   ____

Vowels, Vowels, Everywhere

MATH TOPICS

data collection, analysis, and representation; computation; problem solving

PRIOR KNOWLEDGE NEEDED

1. converting from a fraction to a decimal
2. converting to a percent
3. measuring angles and drawing a pie graph

DIFFERENTIATION STRATEGIES

Principles

Flexible grouping: Students will work in pairs to collect the data and as a whole group for debriefing. Students will create his or her own pie chart.

Ongoing assessment: Observe students and use productive questioning.

Teacher's Strategies

Product: Tiered Journal Questions: Each of the journal questions requires students to employ similar mathematics skills. Journal question Level 1 has fewer variables than journal question Level 2.

According to Students

Learning Styles/Multiple Intelligences: logical/mathematical, verbal/linguistic, visual/spatial, musical/rhythmic, interpersonal

MATERIALS NEEDED

1. a copy of the poem "The Number 5" for each pair of students
2. a data collection sheet for each pair of students; Each student in the pair will need his or her own pie graph.
3. an overhead transparency of the data collection sheet
4. calculators

TEACHING SUGGESTIONS

Engaging the Students

Give each pair of students a copy of the poem "The Number 5" and read it aloud with the group. (The poem can also be sung to the tune of the poem "Trees" by Joyce Kilmer.) Ask the group to estimate which letter or letters they think appear the most often. Ask them to predict which letters will not appear at all (and why). You can refer to the television game show *Wheel of Fortune* and discuss how knowing which letter(s) have the greatest frequency in the English language and which have the least might be helpful for a contestant to know. Explain that students will work in pairs to complete this activity. One student will call out the letters, and the other will tally the number on the frequency table. Explain how to tally by fives as shown here: 𝍸.

The Exploration

Assign students in pairs, and give each group a data collection sheet. Have students collect the data required using the poem. It is necessary for students to work together so that one can call out the letters and the other can tally the results. The table requires the students to tally the frequency of each letter, change the frequency to a fraction and a percentage, and find the corresponding number of degrees in a circle. "Five" is sometimes spelled out and sometimes written as a numeral. The answer key below considers it as 1-f, 1-I, 1-v, and 1-e regardless of how it is written. In addition, all numerals should be translated into words so that each letter can be tallied. The table below shows the number of times and the percentage that each vowel appears.

A	41	6.4%	E	95	14.9%	I	55	8.6%	O	43	6.7%	U	16	2.5%

There are a total of 639 letters in the poem. The percentages have been rounded to one decimal place—the total percent shown as rounded is 39.1% of the total number of letters. According to the book *Cipher Systems*, the frequencies for vowels in the English language are as follows: A: 8.2%; E: 12.7%; I: 7.0%; O: 7.5%; U: 2.8% for a total of 38.2% of all letters.

Debriefing

It appears that there is a slightly higher percentage of vowels in the poem than in the English language. At the conclusion of this investigation, students can be given this data (or research it for themselves on the Internet) and be asked to explain possible reasons for the differences that appear between the two sets of data.

ASSESSMENT

1. Student products: Assess accuracy of frequency table calculations and individual pie charts.
2. Journal questions:
   a. Level 1 question: You are reading a book that contains 120 pages. There are approximately 12 words on each line with an average of 5 letters per word. The frequency of E's in the English language is about 13%. How many E's would you expect to find in the book you are reading?

   b. Level 2 question: You are reading a book that contains 242 pages. There are approximately 12 words per line with an average of 5.3 letters per word. If the frequency of E's in the English language is approximately 12.7%, how many E's would you expect to find in the book you are reading?

VARIATIONS FOR DIFFERENTIATION—TIERING

Students can tally all of the letters of the poem and duplicate the investigation using the same format. The table shown below indicates the number of times each letter appears and the percentage that number represents. The total percentage is less than 100% because of rounding.

A	41	6.4%	B	5	0.8%	C	12	1.9%	D	14	2.2%	E	95	14.9%	F	26	4.1%
G	13	0.2%	H	25	3.9%	I	55	8.6%	J	3	0.5%	K	4	0.6%	L	30	4.7%
M	16	2.5%	N	50	7.8%	O	43	6.7%	P	10	1.6%	Q	1	0.2%	R	39	6.1%
S	41	6.4%	T	48	7.5%	U	16	2.5%	V	23	3.6%	W	19	3.0%	X	2	0.3%
Y	7	1.1%	Z	1	0.2%												

The Number 5

BY HOPE MARTIN

I think that I shall not contrive
A number lovelier than five;
Halfway between 0 and 10
We see it time and time again.

Our world is full of things in fives
Let's see how 5 is in our lives;
We have five senses, don't you know?
Now watch our list just grow and grow!

Geometry is next to see
With pentagons all fives agree,
Plato's five solids were well planned
Pentominoes (5 squares) are grand.

The number 5 is just so prime
The 5¢ nickel is $\frac{1}{2}$ a dime,
Five digits on one foot or hand
Five Olympic rings were planned.

A Fibonacci number—5
Count five from the start at five arrive,
Five-fold symmetry in apple's core
Just look in nature for more and more.

There are five Chinese elements
You can research what they represent,
English language has five vowels
That's it! I throw in the towel!

Vowels, Vowels, Everywhere

Directions: We have just read the poem "The Number 5"—a poem that is really involved with numbers and math! In honor of this poem, let's conduct an experiment and investigate the frequency of times each vowel appears in the poem. Use the table to help you conduct the experiment and analyze the results.

Letter	Tally	Frequency	Fraction	Percentage	Degrees in a 360° circle
A					
E					
I					
O					
U					
TOTAL					

Now that you have collected your data, use it to design a pie chart on the next page.

Vowels, Vowels, Everywhere

Designers: ____

Title of Graph: ____

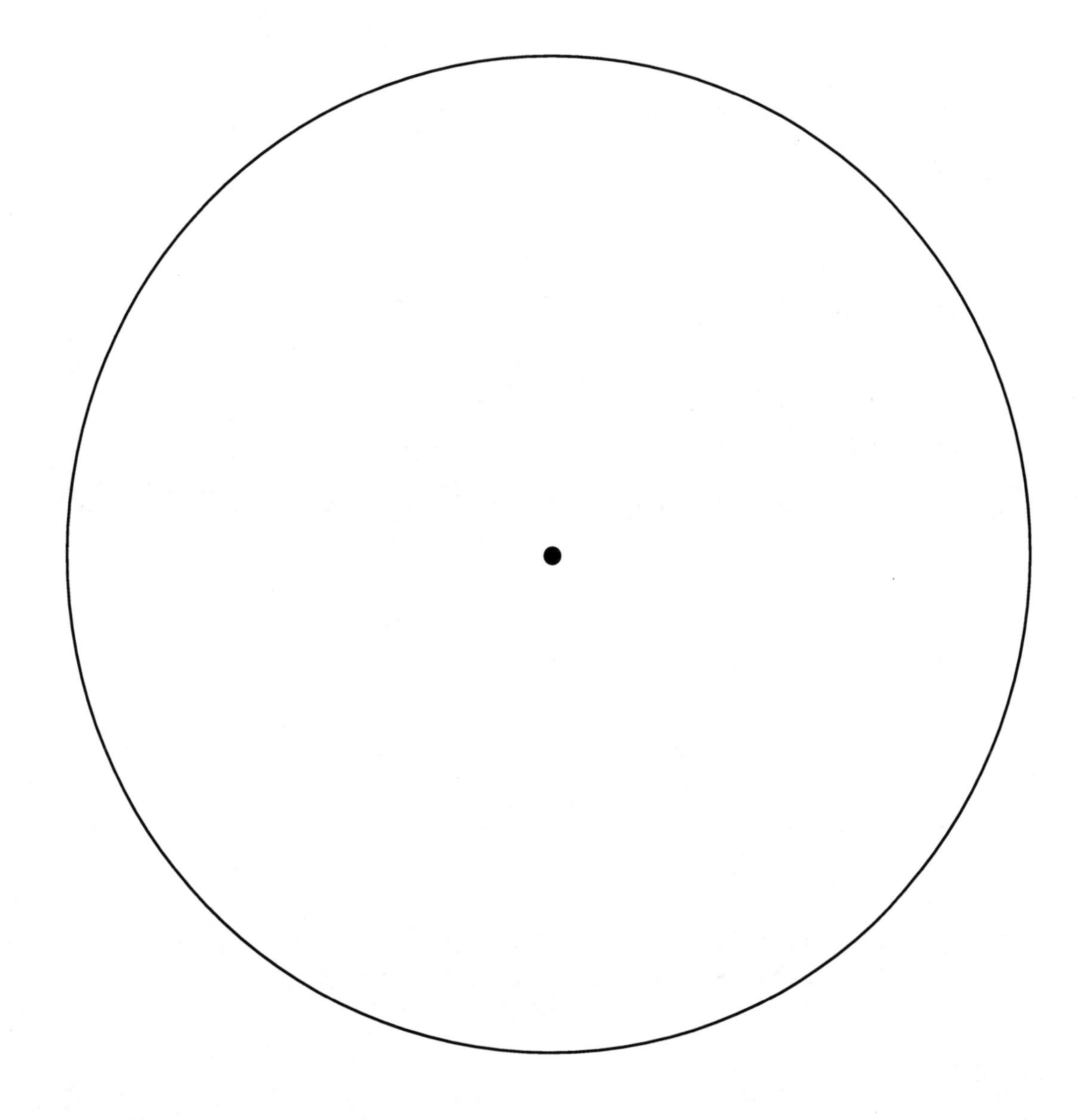

Phone Home

MATH TOPICS

data collection and analysis, measurement, conversions, computation, problem solving

PRIOR KNOWLEDGE NEEDED

1. operating and reading a stopwatch
2. finding averages (means)

DIFFERENTIATION STRATEGIES

Principles

Flexible grouping: Students work in groups of four to conduct the experiment and return to whole-group instruction for debriefing and analysis of data.

Ongoing assessment: Group data sheets that include the problem-solving process can be used as part of the assessment process. Observation by the teacher and the use of productive questions can be used to focus attention, help with measurement, and encourage reasoning strategies.

Teacher's Strategies

Product: Tiered Journal Questions: The Level 1 question is a different version of the same activity students did in class; however, this problem uses Mercury as the location. The Level 2 question requires conversion from miles per second to miles per hour.

According to Students

Learning Styles/Multiple Intelligences: logical/mathematical, verbal/linguistic, bodily/kinesthetic, interpersonal, intrapersonal

MATERIALS NEEDED

1. one stopwatch for each group of four students
2. a student activity sheet and poem "Phone Home" (page 111)
3. clipboards
4. calculators

TEACHING SUGGESTIONS

Engaging the Students

Discuss both the mathematics and science concepts connected to the lesson. When we talk on the telephone, there appears to be virtually no delay between the time we speak and when the person on the other end hears our words. But there is a delay. Because the sound travels over the phone lines at the speed of light (186,000 miles per second), we are not aware of the lag in time.

But what if the phone call is being made to a much more distant place? How long would it take then? When interviews (shown on television) are bounced off satellites, we do experience the slight delay that results because of the vast distances. Ask students how we are aware of this delay.

The Exploration

Miranda is an alien from Mars, and she is sending a message-poem home. Students, working in groups of four, take turns reading the poem and recording how long it takes to read (to the nearest $\frac{1}{10}$ of a second). The group's average is used to calculate how long it would take for the message to reach Mars.

Debriefing

On the board, write the average times for each of the groups. Ask students to analyze these times—are they consistent, or does there appear to be an outlier? The average time for the class can be calculated.

ASSESSMENT

1. Student products: Assess the accuracy of the data collected by student groups.
2. Journal questions:
    a. Level 1 question: How long would it take for your group to send the message to Mercury which is an average distance of about 51,000,000 miles from Earth?
    b. Level 2 question: Sound travels at a speed of 5 miles/second. Mars is 33.6 million miles from Earth. How long would it take for a phone call to reach Mars at this speed? (Be sure to express your answer in reasonable units.)

Phone Home

Directions: We have a visitor from space! Miranda from Mars is really enjoying her visit but is getting a little homesick, so she is making an intergalactic call home.

- Mars is 33,600,000 miles from Earth.
- Telephone talk travels at the speed of light, which is 186,000 miles per second.

How long will it take for Miranda's message to reach Mars?

Work with your group and

1. take turns reading Miranda's message poem using a normal rate of reading speed,
2. time each person using a stopwatch and record his or her time on the table,
3. find the average time it took your group to read the message,
4. use that average to calculate how long it took to "phone home."

Name	Time to Read Message
Average Time	

We calculate the average time it will take for the message to reach home will be ________. These are our calculations:

Phone Home

This is Miranda's message:

Having a wonderful time,
Wish you were here,
Earth is very different,
But really quite near!

I'm sending this message,
On Tuesday, the 10th,
Please check the computer,
How much time has been spent?

We'll all work together,
To figure this out,
How long to phone home,
That's what this is about.

Use a stopwatch—keep time,
How long will it take,
For my family in space,
This rhyme to partake?

How Long Is Your First Name?

MATH TOPICS

data collection and analysis, statistics, absolute value, deviation from the mean, problem solving

PRIOR KNOWLEDGE NEEDED

1. finding the mean or average
2. calculating absolute values

DIFFERENTIATION STRATEGIES

Principles

Flexible grouping: For this activity, students work as a whole group.

Ongoing Assessment

The journal question can be used as part of the assessment process.

Teacher's Strategies

Product: Journal question: This journal question gives students time to reflect on the activity and explain it in their own words.

According to Students

Learning Styles/Multiple Intelligences: logical/mathematical, verbal/linguistic, bodily/kinesthetic, interpersonal, intrapersonal

MATERIALS NEEDED

1. strips of 1-×-1-inch squares (one for each student)
2. a flat surface on which to spread out the names (one under the other)
3. scissors to cut out letters and "even out" the names
4. calculators
5. an overhead transparency of student data sheet

TEACHING SUGGESTIONS

Engaging the Students

Give students a copy of the student activity sheet, and discuss the questions posed on the first page.

The Exploration

Give each student a strip of 1-×-1-inch squares, and have students write their first names in the squares (one letter in each square). Place the strips of names on a table that is large enough to accommodate them (one underneath the other). Ask students the reasoning question "By looking at the strips of names, can you predict the average (or mean) number of letters in our names?" Be sure the students have an opportunity to estimate the mean.

Once students have made their estimates, explain to them that the mean can be approximated by "evening out" the strips. In other words, we can cut letters from the longer names and add these letters to the shorter names to find the mean. But first ask students to find the sum of all of the letters on the table. Now start cutting letters from the longer names and adding them to the shorter names until the columns are even. But what if the columns are not completely even? Ask students between which two numbers the mean will be found. Have them approximate whether the mean will be closer to one of the numbers or the other and why.

Debriefing

Once students have evened out the column of names, ask them how they would normally have found the average number of letters in the names of students in the class. If they say to add up the numbers and then divide, go back to the sum the students found in the hands-on activity. Ask, "How can we use this sum to help us find the exact average or mean?"

ASSESSMENT

Journal question: How did we approximate the mean or average of the class when we "evened out" the names on the table? In your own words, tell how this explains the following statistic: There is an average of 2.5 children per household in the United States.

VARIATION FOR DIFFERENTIATION—TIERING

The activity sheet that asks students to design a form to collect information based upon the length of their names can be used as a whole-class activity or as an enrichment activity for more capable students. Students can be paired homogeneously for this activity.

How Long Is Your First Name?

Directions: How many letters do you have in your first name? Is it a short name or a long name? When you fill out a form that asks you to put the letters of your first name in little boxes, do you run out of boxes or are there some boxes left over? How do you think the designers of these forms know how many boxes they will need so that most people will have enough space to fit in their whole name? Let's conduct two experiments to see if statistics can answer this question.

1. Use a strip of one-inch squares to write your first name. Imagine you are filling out a form and putting one letter in each of the squares.

2. Place your name on a table underneath the names of the other members of the class.

3. Predict the average number of letters in the first names of the students in your class.

4. Pay close attention as the teacher physically "evens out" the names by cutting letters off the longer names and adding these letters to the shorter names.

The average, or mean number, of letters in the first names of the students in our class is ________________.

How Long Is Your First Name?

Name of Student	Number of Letters in First Name	Class Average	Difference (Deviation from the mean)
Average			

How Long Is Your First Name?

Let's use the data we collected and the statistics we calculated to design a form with enough spaces to accommodate most of the first names of people in the United States. Work with your group to answer these questions and design the form.

1. What was the class mean? ______________
2. How many more letters than the mean did the name with the most number of letters have? ______________
3. How many times larger than the mean is this name? (Round to the nearest whole number.) ______________
4. Problem-solve the number of spaces you think the form would need to accommodate the number of letters in the first names of most of the people in the United States. On the lines below, explain your thinking.

____

____

____

____

____

____

____

____

____

____

____

____

____

____

____

Geometric Probability: Dartboards and Spinners

MATH TOPICS

geometric probability, data collection and analysis, fraction and decimal concepts, angle measurements, pie graphs

PRIOR KNOWLEDGE NEEDED

1. addition and multiplication of fractions
2. introductory concepts of probability and "fairness"
3. measuring with a protractor
4. drawing a pie graph
5. converting from fractions to percents

DIFFERENTIATION STRATEGIES

Principles

Flexible grouping: Introductory material and review is accomplished as whole-group instruction; students work in pairs to conduct experiments with a spinner and to design a fair dartboard. Individual students should complete the journal page used to explain the experiment.

Ongoing Assessment: Observe students during the experiment and ask attention-focusing, problem-posing, and reasoning questions of student pairs. Assess the quality of data collection and fair game board, and completeness of journal question response.

Teacher's Strategies

Product: Tiered Journal Questions: Journal question Level 1 is an extension of the activity; the Level 2 question asks students to apply probability concepts to a real gambling situation.

According to Students

Learning Styles/Multiple Intelligences: logical/mathematical, verbal/linguistic, visual/spatial, bodily/kinesthetic, intrapersonal, interpersonal

MATERIALS NEEDED

1. rulers, compasses, and protractors
2. colored markers or pencils
3. paper clips and pencils for spinners
4. calculators
5. overhead transparency of class data table

TEACHING SUGGESTIONS

Engaging the Students

Ask students, "What do we mean when we say that a game is fair or unfair? Is it fair if we win and unfair if we lose?" Give students time to discuss the concept of fairness, place them in pairs, and then pass out the first activity sheet. Read the introductory material with students, and ask them to make a prediction whether the dartboard is fair or not. The fractional parts are as follows:

P(Red)= $\frac{3}{18}$ $\left[\frac{1}{6}\left(\frac{1}{3}\right)+\frac{1}{3}\left(\frac{1}{3}\right)\right]$

P(Yellow) = $\frac{6}{18}$ $\left[\frac{1}{6}\left(\frac{1}{3}\right)+\frac{1}{3}\left(\frac{1}{3}\right)+\frac{1}{2}\left(\frac{1}{3}\right)\right]$

P(Green) = $\frac{5}{18}$ $\left[\frac{1}{2}\left(\frac{1}{3}\right)+\frac{1}{3}\left(\frac{1}{3}\right)\right]$

P(Blue) = $\frac{4}{18}$ $\left[\frac{1}{6}\left(\frac{1}{3}\right)+\frac{1}{2}\left(\frac{1}{3}\right)\right]$

TOTAL $\frac{18}{18}$ 1=100%

Based on these data, the dartboard is not a fair board.

The Exploration

The first problem for the groups of students is to determine why this dartboard is not fair by calculating the fractional probabilities for each of the colored areas. Later in the activity, they will be asked to design a dartboard that is a "fair game."

The group's next task is to design a spinner (pie graph) in which the color segments of the graph represent the same percentage of the circle as they do on the unfair dartboard. Based on the theoretical probabilities of each of the colors, the sections of the pie graph should be red: 60°; yellow: 120°; green: 100°; and blue: 80°. By spinning a paper clip around a pencil, students find the experimental probability of landing on each of the colored segments. When the class data are combined, an experimental probability is calculated for each of the segments of the circle. These data are compared with the theoretical probabilities computed from the dartboard.

Before pairs of students design a "fair" dartboard, they should individually complete the journal question page that is part of the lesson. This should help them with their design.

Debriefing

When each group has designed its own fair dartboard, the dartboards can be shared with the whole class. Copies can be made so students can compute the fractional parts of each of the colored sections to determine if each of the boards are, in fact, fair.

ASSESSMENT

1. Student products: Student activity sheets and journal questions can be used as part of the assessment process; fair dartboards can also be used.
2. Journal questions:
   a. Level 1 question: This dartboard was designed by one of your classmates. Analyze the sections. Calculate what fractional part each of the colors represents. Is this dartboard fair? What can you do to make it fair? 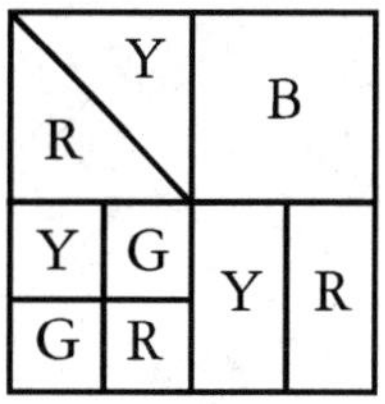
   b. Level 2 question: A sign in a gambling casino boasts that "This slot machine pays a winner 90% of the time." Explain why the longer a player plays, the more money he or she will lose."

Geometric Probability: Dartboards and Spinners

Directions: The dartboard below is in your favorite pizza parlor. Whenever you and your family go there for dinner, you all enjoy trying to get the most points. The points are based upon the probability of the dart landing in a certain area. There are four colors on the board: red, blue, green, and yellow.

If four different people each choose one of the colors, do you think each person has the same chance of winning? Why or why not? ______________________

____

To find the theoretical probability of landing on each of the colors, find the fractional part that each of the colors represents on the board. *Hint*: The yellow in the middle column is $\frac{1}{3}$ of $\frac{1}{3}$ or $\frac{1}{9}$ of the board. Find the totals for each color. Write your answers in the spaces provided.

P(Red) = __________

P(Yellow) = __________

P(Green) = __________

P(Blue) = __________

Does each of the players have an equal chance of landing on his or her color? Do you think this game is fair? Why or why not? ______________________

____

____

____

____

Geometric Probability: Dartboards and Spinners

Directions: You want to play the game of chance at home but you don't have a dartboard. Work with your partner to design a spinner that is the same as the dartboard in the restaurant. Make sure that each of the segments represents the same fractional part of the circle that each of the colors represents on the dartboard. Use the circle below to measure each of the four segments.

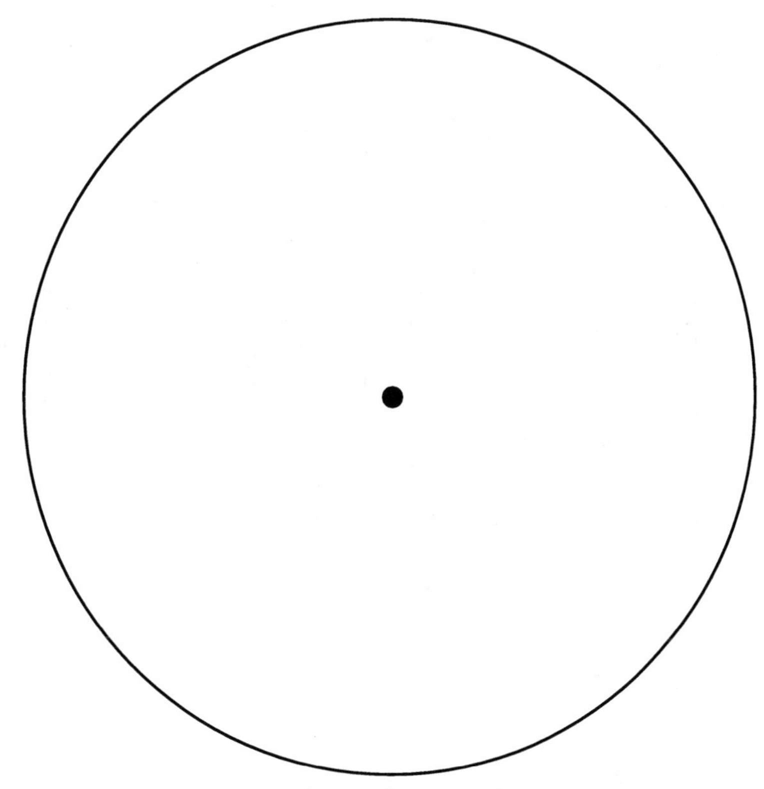

With your partner, use a paper clip and a pencil to play this game. Each player should spin the paper clip at least 20 times. Keep track of where the spinner lands and complete the table. Be prepared to share your results with the other players in the class on the Class Results table.

Color	Tally	Frequency	Fraction	Percent
Red				
Yellow				
Green				
Blue				

Geometric Probability: Dartboards and Spinners

CLASS RESULTS

Group #	Percent of			
	Red	Yellow	Green	Blue
1				
2				
3				
4				
5				
6				
7				
8				
9				
10				
Average				

This is the original dartboard from the pizza parlor. Complete the table—it will help you analyze the experiment.

Colors	My Results (%)	Class Results (%)	Theoretical Probability (%)
P(Red)			
P(Yellow)			
P(Green)			
P(Blue)			

How close were the theoretical and experimental probabilities when the average class results are used?

Geometric Probability: Dartboards and Spinners

JOURNAL PAGE

Use this page to discuss the results of the experiment. You may discuss other things, but be sure to answer these questions:

- What was the probability of landing in each area?
- What fractional part of the circle was each of the colors?
- How many degrees were in each segment?
- Did your group obtain the expected results? Why or why not?
- Was the experimental probability close to the theoretical probability in your group's experiment?
- Was the experimental probability close to the theoretical probability when the class results were recorded?

Geometric Probability: Dartboards and Spinners

OUR DARTBOARD

Directions: Work with your partner to design a dartboard where the geometric probability for each of the four colors is the same. Be sure to use four colors. Draw your picture in the square below. You will need to use a ruler and measure carefully because it has not been divided into sections. Use your imagination and design a dartboard that is interesting to you. Just be sure to make it a fair game.

Flipping Three Coins: Heads or Tails

MATH TOPICS

experimental and theoretical probability, fraction and percent concepts, statistics and data collection, problem solving

PRIOR KNOWLEDGE NEEDED

1. having basic knowledge of theoretical and experimental probability
2. converting fractions to percent

DIFFERENTIATION STRATEGIES

Principles

Flexible grouping: This begins as a whole-group activity and then students work in pairs to conduct the experiment and analyze their results. Finally, during debriefing, student pairs share their games with the whole group.

Ongoing assessment: Observe students during the experiment, and assess understanding through productive questioning and accuracy of data collection sheet prepared by students.

Teacher's Strategies

Product: Journal question: Students' explanation of fairness indicates their understanding of probability concepts.

According to Students

Learning Styles/Multiple Intelligences: logical/mathematical, verbal/linguistic, bodily/kinesthetic, visual/spatial, interpersonal

MATERIALS NEEDED

1. three coins per group
2. one student data collection sheet for each group of students
3. one small paper cup for each group
4. an overhead transparency of student data sheet and class data sheet

TEACHING SUGGESTIONS

Engaging the Students

Assign students in groups of four, and give each group a copy of the data collection sheet. Read the opening paragraphs with them, and lead the discussion with the question, "Do you think this game is fair?" Give students an opportunity to discuss the question, and then say, "Let's see if this game is fair. You are going to conduct an experiment with your partners and keep track of the results on the frequency table." Demonstrate for the class how to record the results of each coin toss by using tally marks. When they reach five, the tally should look like this: ~~||||~~.

Explain that when each group has completed the activity, they are to write their results on the class data table for analysis.

The Exploration

After the initial discussion, read the rules of the game. Have students choose whether they want to receive points for 3H, 2H-1T, 2T-1H, or 3T. During this time, the teacher can walk around and ask attention-focusing questions, such as, "Does each of the possibilities appear to be occurring at an equal rate?" or "What have you noticed about three heads or three tails?"

Debriefing

This part of the lesson will focus on the results of each of the groups and the analysis of the results. When the average, or mean, of all of the groups is calculated, the results will indicate that the game was not fair. This table shows the sample space and probabilities:

<table>
<tr><th>First Coin</th><th>Second Coin</th><th>Third Coin</th><th>Combination of 3 coins</th><th>Theoretical Probabilities</th><th>Percent</th></tr>
<tr><td rowspan="4">H</td><td rowspan="2">H</td><td>H</td><td>HHH</td><td rowspan="2">3 H = $\frac{1}{8}$</td><td rowspan="2">12.5%</td></tr>
<tr><td>T</td><td>HHT</td></tr>
<tr><td rowspan="2">T</td><td>H</td><td>HTH</td><td rowspan="2">2H 1T = $\frac{3}{8}$</td><td rowspan="2">37.5%</td></tr>
<tr><td>T</td><td>HTT</td></tr>
<tr><td rowspan="4">T</td><td rowspan="2">H</td><td>H</td><td>THH</td><td rowspan="2">2T 1H = $\frac{3}{8}$</td><td rowspan="2">37.5%</td></tr>
<tr><td>T</td><td>THT</td></tr>
<tr><td rowspan="2">T</td><td>H</td><td>TTH</td><td rowspan="2">3 T = $\frac{1}{8}$</td><td rowspan="2">12.5%</td></tr>
<tr><td>T</td><td>TTT</td></tr>
</table>

ASSESSMENT

1. Student products: Assess accuracy of data sheet and quality of "fair game" developed by group.
2. Journal question: Explain how you determined if the game was fair. How did you alter the rules so that the game would be fair to all four players?

VARIATIONS FOR DIFFERENTIATION—TIERING

Students can duplicate the experiment using two coins and then four coins. Encourage students to find the pattern that develops; the number of possible occurrences is related to the number of coins used. If n is equal to the number of coins, then the number of potential outcomes in the sample space is equal to 2^n.

Flipping Three Coins: Heads or Tails

Suppose you are playing a game with three of your friends. These are the rules of the game:

1. Each of the four players chooses one of the following: HHH (meaning that all three coins land on heads), HHT (two of the coins land on heads and one of them lands on tails), TTH (two of the coins land on tails and one lands on heads) and TTT (all three coins land on tails).
2. When the coins land on your choice, you get one point.
3. The person who has the most points at the end of 40 tosses is the winner.

Do you think this game is fair? Let's talk about it.

Directions: Now let's play the game to test our thinking! In your group of four, place three coins in a small paper cup, shake the cup well, and carefully turn it over. Record the occurrence. If the coins were all heads (HHH), place a tally mark in that row. Complete the frequency table, and be prepared to share your results with the rest of the class on the class data table.

Results of Toss	Tally	Frequency	Fraction	Percent
HHH				
HHT				
TTH				
TTT				

Does this appear to be a "fair game"? Did each of the players have an equally likely chance of winning? Explain your reasoning.

____

____

____

____

____

Share your results with the other groups—use the class data table.

Flipping Three Coins: Heads or Tails

CLASS DATA

Group #	Percent of			
	HHH	HHT	TTH	TTT
1				
2				
3				
4				
5				
6				
7				
8				
9				
10				
Average				

After analyzing the class data, rewrite the rules of this game so that it becomes a "fair game." Be prepared to share your game with the rest of the class.

Flipping Three Coins: Heads or Tails

Use this page to explain how you would make this a fair game.

Brush Up Those Skills
Chapter 4

Directions: Roll one die; your group will work on the activity that matches the number on the die. If you have completed that activity, roll the die until it lands on an activity you have not done. On the next page, check off the activity your group completed. Make sure each group member has signed his or her name.

1	Work with your group to design a survey. Follow the steps below. 1. Carefully write the question you wish to ask. 2. Choose four or five answer choices you will give to students. 3. Design a frequency table to collect the data containing the choices, frequency each occurred, fraction it occurred, and percent each occurred. 4. Design a graph to represent the data. It can be a bar or pie graph. 5. Write an analysis of your survey as if it were an article for a newspaper. What was your question? Who did you ask? What were the results?
2	Use a protractor and a compass to design a spinner in which 1/4 is red, 1/3 is blue, 1/3 is green, and 1/12 is orange. If you were to spin the dial 36 times, how many times would you expect each of these colors to appear? Conduct the experiment and record your experimental results on a frequency table. Now compare your results to theoretical probability.
3	Ask 50 students their month of birth. Design a frequency table with these columns: Month, Tally, Frequency, Fraction, Percent. Analyze the data and answer the following: 1) the month with the fewest births 2) the month with the greatest births. Ask students born in the same month for the date of their birth. Did any have the same birthday? If yes, how many? Write a newspaper article answering who, what, where, and why questions.
4	Choose a paragraph from a newspaper article. Design a frequency table showing the number of letters and vowels in each word. Organize your data and graph the results. Label the vertical axis "Number of Vowels," and the horizontal axis "Number of Letters." With your group, describe the relationship between the number of letters and vowels in a word.
5	C A R D Some students are going to spin a spinner and then draw one of these cards from a bag. Draw a two-stage, eight-branch tree. List all the possible outcomes. Express them in fraction form.
6	How often will a sum of 7 appear when we roll 2 dice? With your group, conduct an experiment to answer this question. In a table, record your results with these headings: Trial, Tally, Frequency. Conduct 10 trials. For each one, roll a pair of dice until you obtain a sum of 7. Tally all your trials. Write the number of trials it took before a 7 appeared. Conduct this 10 times. Find the mean or average number of trials to reach a sum of 7. Analyze your results and explain how your results compare to the probability of rolling a sum of 7.

Brush Up Those Skills
Chapter 4

Our Progress Completing These Activities

We have completed | **Signatures of Group Members**

☐ Activity 1 ______________________

____

____

____

☐ Activity 2 ______________________

____

____

____

☐ Activity 3 ______________________

____

____

____

☐ Activity 4 ______________________

____

____

____

☐ Activity 5 ______________________

____

____

____

☐ Activity 6 ______________________

____

____

____

Selected Answers

Page 20—Baking Blueberry Muffins

A. 1. 96 muffins

2. \$0.375 each or $37\frac{1}{2}$¢ each

3. 300% profit

B. 1. 57 muffins (there is a remainder but that is not enough to make another muffin)

2. \$0.64 each (although the remainder is less than $\frac{1}{2}$, it is added on to calculate cost)

3. \$30.72/dozen (students may round up the number to a more reasonable amount)

Page 57—Pattern Block Patterns

The fifth term will have 4 triangles and 5 trapezoids.

Since each term has t-1 # of triangles and t # of trapezoids, the 20th term will have 19 triangles and 20 trapezoids; the hundredth term will have 99 triangles and 100 trapezoids; the nth term will have n–1 triangles and n trapezoids.

Page 65—Pennies and the Sears Tower

The number of pennies needed to reach a height of 1450 feet will depend on the student-explorations. The students' calculations will depend on the age of their pennies, the number of pennies they used to make their estimates, and the accuracy of their measurements. Because of all of these variables, it is necessary to collect the data from all of the groups and analyze the individual results.

Page 106—Vowels, Vowels, Everywhere

The vowel A appears in the poem 41 times or 6.4% of the total # of letters (639).

The vowel E appears in the poem 95 times or 14.9% of the total # of letters (639).

The vowel I appears in the poem 55 times or 8.6% of the total # of letters (639).

The vowel O appears in the poem 43 times or 6.7% of the total # of letters (639).

The vowel U appears in the poem 16 times or 2.5% of the total # of letters (639).

Vowels	Frequency	Fraction	Percentage	# of ° in circle
A	41	41/250	16.4%	59°
E	95	95/250	38.0%	137°
I	55	55/250	22.0%	79°
O	43	43/250	17.2%	62°
U	16	16/250	6.4%	23°
TOTAL	250	250/250 = 1	100.0%	360°

Title ____________________

Teacher's Page

MATH TOPICS

PRIOR KNOWLEDGE NEEDED

1.

2.

DIFFERENTIATION STRATEGIES

Principles

Flexible grouping:

Ongoing Assessment:

Teacher's Strategies

Product: Tiered Journal Questions:

According to Students

Learning Styles/Multiple Intelligences:

MATERIALS NEEDED

1.

2.

3.

4.

TEACHING SUGGESTIONS

Engaging the Students:

The Exploration:

Debriefing:

ASSESSMENT

1. Student products:
2. Journal questions:
    a. Level 1 question:
    b. Level 2 question:

Brush Up Those Skills

Skill ______________________________

Directions: Roll one die; your group will work on the activity that matches the number on the die. If you have completed that activity, roll the die until it lands on an activity you have not done. On the next page, check off the activity your group completed. Make sure each group member has signed his or her name.

1	
2	
3	
4	
5	
6	

Brush Up Those Skills

Skill ______________________________

Our Progress Completing These Activities

We have completed	Signatures of Group Members
☐ Activity 1	______________________
	______________________
	______________________
	______________________
☐ Activity 2	______________________
	______________________
	______________________
	______________________
☐ Activity 3	______________________
	______________________
	______________________
	______________________
☐ Activity 4	______________________
	______________________
	______________________
	______________________
☐ Activity 5	______________________
	______________________
	______________________
	______________________
☐ Activity 6	______________________
	______________________
	______________________
	______________________

TAP Activities

Topic(s) ______________________________

TASK	AUDIENCE	PRODUCT
an original idea presented by student to be approved by the teacher		

Pascal's Triangle

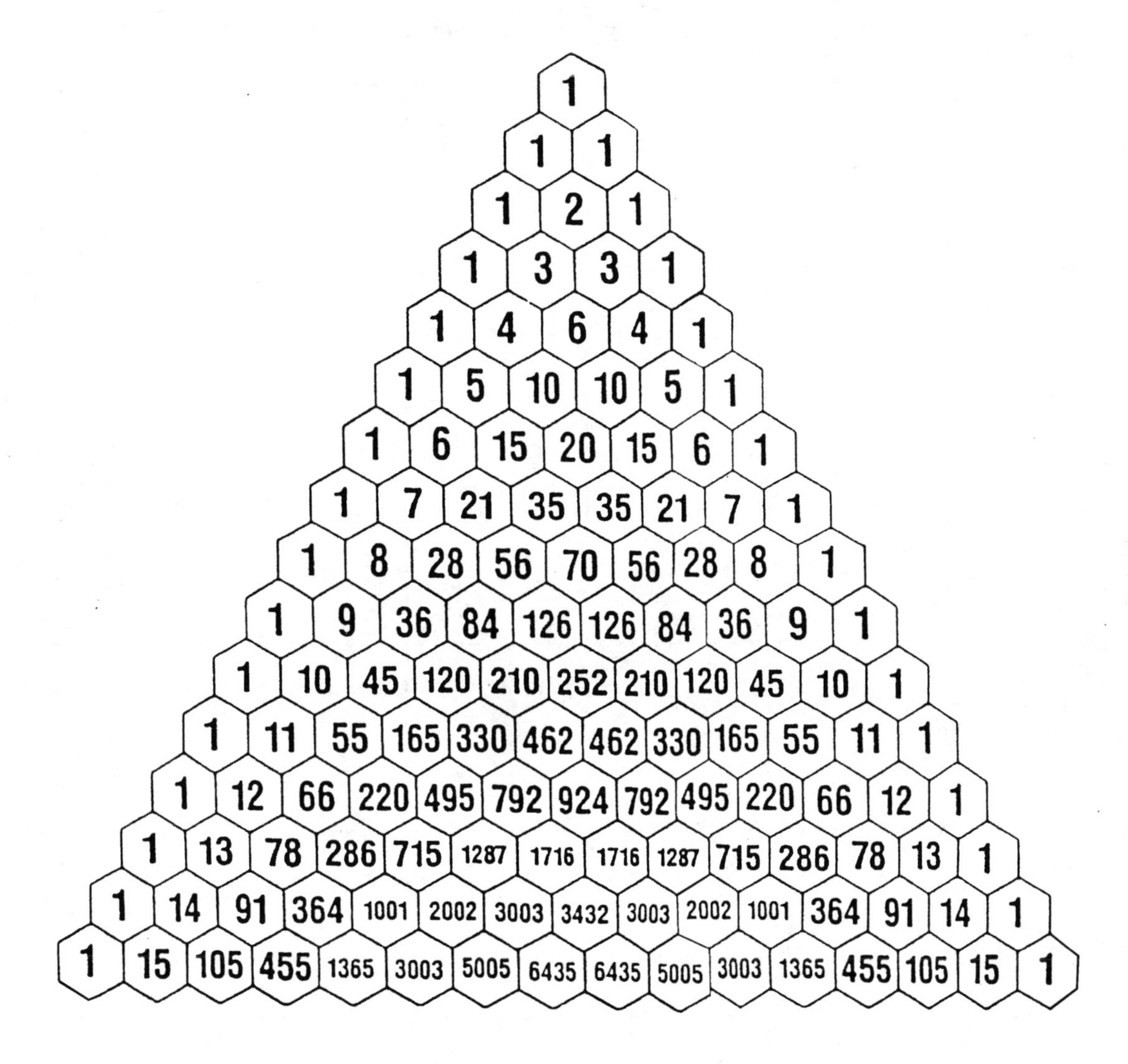

One-Inch Squares

for “Soda Pop Math” and “How Long Is Your First Name?” activities

Bibliography

Armstrong, T. *Multiple Intelligences in the Classroom.* Alexandria, VA: ASCD, 1994.

Beker, H. and F. Piper. *Cipher Systems.* New York: Wiley, 1982.

Countryman, J. *Writing to Learn: Strategies That Work.* Portsmouth, NH: Heinemann, 1992.

Gardner, H. *Frames of Mind.* New York: Basic Books, 1985.

Gee, J.P. "The Classroom of Popular Culture: What Video Games Can Teach Us About Making Students Want to Learn." *The Harvard Education Letter*, Nov/Dec 2005.

Gregory, G.H. and C. Chapman. *Differentiated Instructional Strategies: One Size Doesn't Fit All.* Thousand Oaks, CA: Corwin Press, 2002.

National Research Council. *Adding It Up: Helping Children Learn Mathematics.* J. Kilpatrick, J. Swafford, and B. Findell (Eds.). Washington, DC: National Academy Press, 2001.

National Research Council. *Everybody Counts: A Report to the Nation on the Future of Mathematics Education.* Washington, DC: National Academy Press, 1989.

National Council of Teachers of Mathematics. *Mathematics Assessment: A Practical Handbook for Grades 6–8.* Edited by W.S. Bush and S. Leinwand. Reston, VA: NCTM, 2000.

National Council of Teachers of Mathematics. *Principles and Standards of School Mathematics.* Reston, VA: NCTM, 2000.

Neil, M.S. *Mathematics the Write Way: Activities for Every Elementary Classroom.* Larchmont, NY: Eye on Education, 1996.

Pappas, T. *The Music of Reason.* CA: Pub Group West, 1995.

Tomlinson, C.A. *The Differentiated Classroom: Responding to the Needs of All Learners.* Alexandria, VA: ASCD, 1999.

Tomlinson, C.A. and C.C. Eidson. *Differentiation in Practice: A Resource Guide for Differentiating Curriculum.* Alexandria, VA: ASCD, 2003.